CHICAGO PUBLIC LIBRARY
HAROLD WASHINGTON LIBRARY CENTER
R0017967954

DATE			

D1550389

Energy

The Rude Awakening

Energy

The Rude Awakening

Richard Bailey

Energy Education Publishers

Published in the United States and Canada by Energy Education Publishers, 1432 Wealthy Street, Grand Rapids, Michigan, 49506, by arrangement with McGraw-Hill Book Company (UK) Limited.

Cataloging in Publication Data

Bailey, Richard
Energy.
1. Power resources – Great Britain
I. Title

333·7 HD9502.G72 77-89057

ISBN 0-918998-03-4

1 2 3 4 JWA 7 9 8 7

PRINTED AND BOUND IN GREAT BRITAIN

Contents

Foreword to the American Edition

Comprehension of the free world energy problem is an involved process, and the beginning of that process requires the shedding of many fixed ideas about our political economy. Energy is not a commodity that can be called forth from the bowels of the earth in unlimited quantities through economic manipulation of the price structure, whether we call that manipulation the force of the free market or political intervention. Energy cannot be isolated as a component which can be added to or subtracted from the Gross National Product. Energy is the multiplier of all other components of the economy. Energy is the *sine qua non*.

The origin of the free world energy problem is the Industrial Revolution, the beginning of modern-day economics interpreted by Adam Smith in his thesis on the wealth of nations. Then, man's energy was released to exploit the raw materials of the world without political restriction. It began with the theory and the practice of unlimited economic expansion. It began in Britain.

British perspective on the rapidly developing world energy problem has more than historic advantage. The inevitable consequence of energy attrition is the necessity for central economic planning and Britain has had considerable experience in mixing the free market forces with governmental direction. Britain has specific experience with the planning for and control of its principal energy resource, coal. Britain's experience with the mixed economy has not been free of conflicts in economic and political motivations, but the British experience has been played out with a high degree of respect for the opposing philosophy. Both sides have had time to test their persuasions on the economy and consider the consequences. Britain has been there with both planning and acute energy attrition.

The Englishman is also a more happy fellow as he surveys the immediate future. He is not as far gone as his American cousin in addiction to petroleum and energy consumption. His newly found petroleum reserves are much larger relative to his needs. He is weathering severe economic storms, but the horizon is brighter for him. He has more time to consider his economic future and he has more capacity for judgement.

By contrast, the American, whose decisions are most vital to the free world's energy problem, has not yet awakened to the existence of a problem. His politicians have only just begun to argue. His economists are solidly opposed to any public discussion. He still lives in a world of confidence in the frontier economy, the efficacy of free enterprise, faith in his scientists, and hope that sacrifice is for the poor barbarian who has strayed from the righteous economic path.

Britain and the United States are the major industrial countries with the largest energy reserves available for their own disposition in the immediate future. Their situations are very similar, but Britain is closer to the free world scene. Britain understands the problem. The British perspective is most important to the American student of the problem. Their perspective is more important than the views of his own economists.

The energy problem is presenting the free world with its moment of truth. It is not the free economy that is at stake, it is the free society. The free economy is a principal support of the free society and, in the minds of many, the two are inseparable. But this view must be recognized as a luxury we can no longer afford. A great deal of our free society can be salvaged in a substantially planned and centrally controlled economy, but neither the society nor the economy can function without energy. The sacrifice of unlimited market freedom, the necessity for a reduction in per capita energy consumption, and the probable demise of the growth economy are devastating considerations for the mainline economists of the free world. Resistance to recognition is still building in the United States. America has not yet begun to think in terms of social and political horizons beyond the next election. There is no clear indication as to how their political and economic leaders can bc turned to face a reality that is now obvious to the scientific community.

The conflict between the economic theory of unlimited expansion and the reality of our finite world is building. It is no longer a decade or more away. These are the words that describe the conservative's optimism, the specifics of a resource, or the 'necessity' of social or political adjustment. The time for recognition of the problem is in the past – not a decade in the future. The time for decision is now. We no longer have decades. We do not even have years before social disruption will be measured in the days of deferred judgement.

In my writings on the subject *Energy – a Critical Decision*, I have maintained a broad perspective, the geological projection of remaining resources, the social and political implications, the conflicts between economic theories. This broader focus omits the specifics of the world organizational response to the energy revolution of 1973, the events before and after the Arab boycott, and the coming of age of the Organization of Petroleum Export Countries.

In this present book, Richard Bailey presents this closer focus on the energy problem with the added advantage of the British perspective. He has been a part of these events, adding economic credentials to his background. His book is necessary in completing the energy picture for the student, for the economist and for the politician.

The British could not afford the luxury of unlimited political-economic conflict. They have had years of experience with the mixed economy. We have the same ideological and cultural roots. We, too, are going to experience the economic necessities that controlled their recent history. As a consequence, it is the British experience, the British experimentation with organizational structure and improvisations with the mixed economy, that will be most instructive in finding solutions to the developing American energy problem. Richard Bailey is in the best position to discuss the details of the world organizational response to OPEC for the economists who are advising the free world political decision makers.

The energy problem is a free world problem. The solutions must be found with world perspective.

August
1977

Samuel M. Dix

Preface

One of the few hopeful signs for our future economic well-being is the emergence of the convention requiring those dealing with energy problems to begin with a comment on the fallibility of energy forecasts. Unfortunately, such disclaimers are often followed by statements making it clear that no such reservations apply to the author's own predictions. Here I have tried to sustain an objective attitude, although it has sometimes been hard to prevent pessimism from breaking through.

This book brings together thoughts and ideas I have developed in papers, articles and lectures, particularly at Ashridge Management College, over the past decade or so. In my case, the idea that imported oil would be available indefinitely was dispelled by a period of secondment to prepare the first five-year development plan of an important Middle East oil producing state. The furore over North Sea oil fortified doubts formed during service with the National Economic Development Office on the suitability of our system of adversary politics for solving long-term economic and social problems. What I have tried to do here is to place Britain's energy problems – which are much lighter than those of her EEC partners – in a wider perspective. In energy terms, what happens to the coal industry during the rest of this century is complementary to the development and running down of

North Sea oil and gas. The need to exploit our oil reserves for their contribution to easing the balance of payments deficit, rather than for their fuel content, is a major complication in an already difficult situation. Farther afield there is the impact of the rise of OPEC on both the rich and poor nations to be considered. A shortage of oil some time after 1985 is forecast unless Saudi Arabia and other OPEC states with only a limited need for increased oil revenues decide to raise their level of production. Deeper down are the social and environmental problems arising both from the exploitation of coal and oil reserves, and from the further development of nuclear power.

I should like to express my gratitude to the various individuals and organizations who have helped me to collect and sift the material presented here, without holding them in any way responsible for the use I have made of it.

Westminster
1977

Richard Bailey

Introduction

Over the past two decades Britain has had fuel crises rather than energy policies. The remark attributed to Aneuran Bevan that Britain was 'sitting on coal, surrounded by fish, but otherwise not well provided for', is no longer strictly true. There is still plenty of coal but several of the old coalfields are nearly worked out, leaving economic and social problems behind them. To maintain and increase coal output involves opening up new deposits in areas hitherto unaware of what lies below or ahead of them. Fish are a rapidly declining asset due to over-fishing of the surrounding seas. On the positive side is the discovery of oil and natural gas under the North Sea Continental Shelf. So in the 1970s Britain has not only coal, but also oil and natural gas and nuclear energy know-how, even though herring and mackerel catches are not what they were.

As a four-fuel economy, Britain clearly has advantages over countries dependent on imports of OPEC oil. The build-up to self-sufficiency in oil holds out the prospect of an end to the balance of payments deficit which has been with us since the Second World War. However, this relief, welcome as it is, will last only for the life of the oil and gas reserves. The prospects on the energy front can be regarded in three stages. In the first, which will last until self-sufficiency in oil supplies is reached, oil imports will continue. This stage will be followed by a period in which we have enough for our own use and a surplus for export. In the third stage North Sea oil will be exhausted and we shall be importing oil once more. This pattern is shown in the

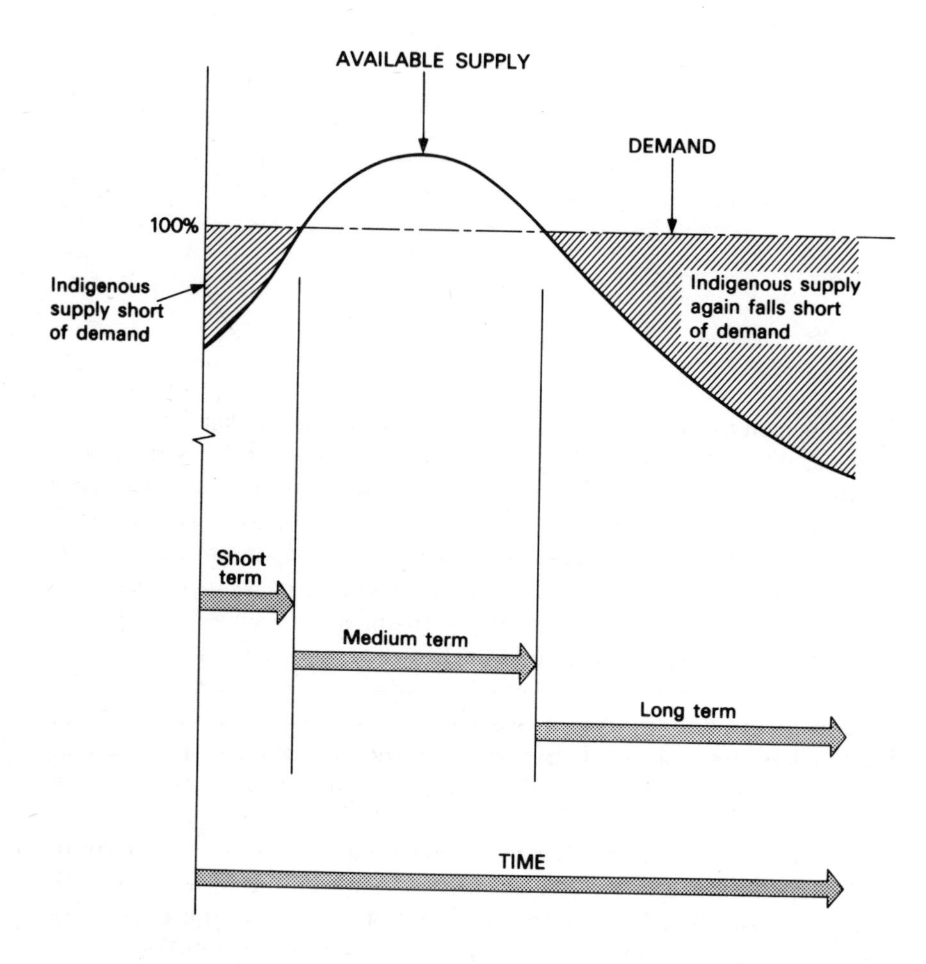

Britain's energy future: Primary energy available from indigenous non-nuclear sources, relative to total demand.
Note. The available supply line implies that Britain could be a net exporter during the medium term. In the event, this will depend on depletion policy and on the contributions from nuclear power and alternative energy sources. (From Energy Paper 11, 'Energy research and development in the United Kingdom', © HMSO. Reproduced by permission.)

diagram. The second stage will be over by the end of the century, on the basis of present knowledge of the extent of our oil resources and the likely rate of depletion.

In the North Sea the drilling rigs have to work in the relatively favourable weather between May and September, the period known as the 'weather window'. For the British economy the second stage, during which oil is being exported, is our 'energy window'; the period of respite before imports of OPEC oil begin once more. The 'weather window' is the period of greatest activity during which key jobs have to be completed. In economic terms, 'the energy window' gives a breathing space during which the restructuring of industry and the eradication of the congenital economic and institutional problems can be undertaken. If this can be done we may emerge into the third stage able to pay for oil imports out of increased exports of manufactured goods. Japan, at present the most successful industrial nation, has no indigenous sources of energy but has managed to make startling inroads into the markets of Britain and other EEC countries. This is the whole point of 'the energy window'. While Britain is unlikely to develop into a facsimile of Japan, there are clearly a great many changes that could be made to cure the economy of its chronic weakness.

At the same time we have to pay attention to developments abroad and their effect on British policies. The prospect of an EEC common energy policy should not, as it sometimes appears to do, give rise to suspicions that EEC members are out to grab a share of our oil. The strengthening of Britain's economic position should enable us to exert greater influence in the EEC, and begin to think in terms of advantages of membership, and possible trade-offs, instead of protesting that the sovereignty of the Westminster Parliament will be undermined.

The oil position gives us an advantage for the medium term, but leaves a lot of unanswered questions about what happens when the oil runs out. For the next 20 or more years we are sheltered from the great energy crisis centred on the stark fact that world supplies are not sufficient to meet total demand. While the world supply of oil is controlled by OPEC, the position of the United States dominates the demand for energy. Although accounting for only 6 per cent of world population, the US consumes 27 per cent of the world's energy. It has been estimated that New York consumes almost as much energy a year as the whole of Africa. This total includes gas, coal and nuclear power as well as oil. President Carter's announcement of a new energy policy in April 1977 recognized the fact that if world demand continues to grow at an annual rate of 5 per cent, known reserves of oil will run out by 1990. The answer clearly is to conserve oil reserves and use more coal, which is abundant, while developing nuclear power, and the as yet unrealized potential of the alternative energy sources from wind, tidal, solar and geo-thermal power. The developing countries are clearly not in a position to make any significant economies as for many of them their consumption is already below the world annual average of 1·5 tonnes of coal equivalent per head.

The imbalance of United States energy consumption is not a new phenomenon. What is new is the fact that increasing demand can no longer be met from indigenous reserves, and the rise in international oil prices makes it impossible to go on importing over 40 per cent of US oil requirements. It is this high dependence on imports which has created the monopoly supply power of OPEC. President Carter's policy of conservation backed up by high taxes to reduce oil consumption, set out to reduce total demand and to lead to a shift from oil to coal as a source of power. However, the production of coal by strip-mining is strongly opposed on environmental grounds, while the extensive use of nuclear power stations, especially the development of the fast breeder reactor, is not acceptable to a wide section of US public opinion.

Part 1 deals with Britain's energy problems, relates them to the state of the economy, and indicates policy options. Part 2 gives an account of the world energy situation and explains how apparently distant events affect our position. In the Epilogue the future is scanned, and an assessment made of the energy prospects in the next century.

Part I
Britain's energy problems

Chapter 1
The hidden agenda

The fortunate isle

Forecasts of varying accuracy and degrees of optimism have been prepared to show that North Sea oil will last until the end of the century, give or take a decade, and that coal reserves are sufficient to last for as long as 300 years. In the background, ready to be updated and brought into service, is Britain's knowledge and experience of nuclear technology. How fortunate for Britain that North Sea oil should have been discovered at this time when imported oil has quadrupled in price, and the greater part of world reserves are under OPEC control. Our partners in the EEC are worrying about the security of imports and pressing forward with costly nuclear programmes, while even the US can no longer be regarded as self-sufficient in fuel. There are considerable grounds for satisfaction in Britain's good fortune in possessing ample supplies of the main fossil fuels. Could these be sufficient to ease the balance of payments position long enough to enable the major structural faults in the British economy to be eradicated? Or are the two sets of problems on economic and energy policies quite separate?

Items on the agenda

Unfortunately the energy problem cannot be dealt with in a vacuum. A whole agenda of related problems is hidden behind the simplest statement of an energy policy for Britain. These concern the actual provision of the various fuels, the uses to which they will be put, and rates at which they will be used up. Because of the peculiar characteristics of the fuel industries, present supplies depend on investment undertaken five, ten or more years ago. By the same token, supplies to the end of the century depend on decisions taken now on the rate of depletion of oil and gas, and on the appropriate related investment in terms of pipelines, terminals, and refineries. Forward investment in power stations has to be planned now and a policy adopted on the fuels they will use. The coal industry already has a major development programme requiring investment phased over the next 20 years. From what is known about the scale of reserves of North Sea oil it is clear that the energy policy agenda must reach forward in time to provide for their inevitable eventual exhaustion.

A further set of problems relate to the level of consumption of total energy and to the mixture of fuels making up the total. This takes the energy problem into the central areas of economic policy and brings in calculations on the impact of growth rates on fuel use. The possibility of exporting oil and increasing the level of coal exports, raises the question of the balance of payments and the effect of exports on the depletion rate and therefore on the date at which self-sufficiency ends and imports begin again.

The high level of investment demanded by the stark conditions of North Sea operation has its repercussions on the national rate of indebtedness. Large sums have been borrowed abroad to help close the yawning gap in the British balance of payments. There is a very real sense in which part of the earnings from North Sea oil exports has already been mortgaged. This is another significant item on the hidden agenda.

International aspects

Yet another problem area is concerned with Britain's membership of the EEC and the possible complications this may raise. A common energy policy might seek to impose limitations on the disposal of British fuel resources, or to give priority to exports to the EEC partners. These questions have not yet arisen but there has been some plain speaking to the effect that Britain cannot expect special treatment over the weakness of sterling and the balance of payments, and at the same time make no concessions over the disposal of her fuel reserves. At this point British energy policy becomes an important item on the agenda of the EEC.

Farther afield, the energy problem becomes entangled in the North–South dialogue between the industrialized countries and the Third World. This relationship differs for the oil exporting countries of OPEC, the developing countries whose earnings from the export of commodities are adequate to pay for their necessary imports, including oil, and the poorest developing countries which have not the means of paying for imports of oil, commodities, foodstuffs and manufactures. The position of the industrialized countries in this dialogue is similarly complicated by the division into those with some indigenous fossil fuels and those entirely dependent on imports of fuel. A further distinction has to be drawn between those like Japan, with a strong trading position which enables them to pay for fuel imports, and those for which the absence of indigenous resources of energy is an embarrassment. Yet another division is between industrialized countries which have built up a considerable nuclear power capacity and those which have scarcely begun to do so.

Attitudes and opinions

As if these varied aspects of the energy problem were not enough, a division of attitudes and opinions exists within each country, industrialized or developing, on the effect of industrial advance on the pollution of the atmosphere and the menace of the self-inflicted devastation of the environment. In the Third World, industrialization is seen as a threat to traditional values and ways of life. In the West, where industrialization is well established, the change of emphasis from production to consumption, with welfare organized by the state in favour of particular groups rather than society as a whole, has solved some old problems but created new ones of an even greater complexity. Among these is the problem of the vast public bureaucracies which between them control large sectors of the economy, including the energy industries. To some extent the movement for state ownership is the result of the slow development of the market system. But in recent years at least, there is no doubt that high company taxation and the low productivity of capital have almost destroyed even the admittedly imperfect free market system that did exist.

The effect of the 1973 oil crisis on the British market economy has been extremely serious. Circumstances and not, as formerly, ideological preferences have shifted the control of development increasingly to the public sector. With the coal, gas, electricity and nuclear industries under public ownership, and state control now being imposed on the oil industry, the fuel industries are at the centre of this change of direction. Circumstances are against those who see themselves as guardians of the private enterprise system, in the context of an energy policy. The profit motive has been stifled, and replaced by official bodies such as the National Enterprise Board. These are still early days for deciding what the new type economic policies – based on resolutions passed at conferences by those who will not have to decide how they can possibly be translated into action – will produce. However, it is not too soon to observe that state control over the 'levers of power' has led to the creation of bureaucracies which are themselves interest groups influencing the scope and operation of the policies which governments can adopt. A change in direction, with power within the government machine shifting from Parliament to the Cabinet, with the bureaucracies and the trade unions exerting increasing influence, has been developing since 1945, but has appeared in more active form since the 1974 elections.

The rude awakening

The name of the international organization responsible for bringing the energy problem up to a leading place in governmental priorities is now firmly in the public mind. Whatever OPEC may do in the future, it is highly unlikely that it will ever again succeed in capturing the undivided attention of the British people as it did with its announcement of the doubling of oil prices and the placing of selective embargoes on supplies on 14 October 1973. Up to that time, oil was generally regarded as a commodity sold by giant multi-national corporations whose brand names were household words. In Britain, 52 million people, 12 million cars, were served by no fewer than 32 000 filling stations. The idea that all those pumps could gradually dry up, diesel locomotives grind to a halt and the radiators of the oil-fired home central heating systems grow cold, produced a wave of apprehension. For the first time, people realized that oil supplies could not be taken for granted. In the first days of surprise and uncertainty motorists rushed to the petrol stations to get their tanks filled up. A Member of Parliament who tried to jump a queue with his car was fined for obstruction. The sale of petrol in cans was forbidden. Ministers (junior), civil servants (higher), business executives, and even some Lord Mayors were issued with smaller cars for their public use. OPEC had succeeded in doing what the energy experts, conservationists, economists and the rest of the professional forecasters had all failed to accomplish. It had brought the British public face to face with the simple truth that supplies of natural resources are not inexhaustible and that the availability of cheap oil in the 'fifties and 'sixties had been a temporary phase in the development of producer power.

Action and response

The adjustments in attitudes towards energy policy since October 1973 have all centred on the new situation in which the supply and price of oil are determined by the producing countries banded together in OPEC. As will be seen later, the move towards greater participation in the production of oil, and control over its distribution by the producing nations, had been developing through the 'sixties. The outbreak of the Arab–Israeli war of October 1973 brought what had been a gradual process to a rapid conclusion, the repercussions of which are still being felt throughout the world. Not surprisingly, the OPEC action did not produce a united or coherent response from the industrialized countries, but a rather confused series of individual exercises by governments aimed at securing oil supplies for themselves. This was true of the EEC, where no immediate steps were taken to help the Federal Republic and the Netherlands, both of which were under embargo. Instead, ministers from Britain, France and other countries hurried to the Middle East to attempt to make the best arrangements they could to secure national supplies. Since then, attempts to secure greater co-operation between oil importers and producers have followed each other in rapid succession, in the UN, in the newly formed International Energy Agency, and in the EEC.

At the same time, the fuel importing countries have made considerable efforts to increase production of indigenous fuel supplies. In the case of Britain this has meant concentration on boosting the rate of investment in North Sea oil and raising the output of coal. At the global level the emergence of OPEC as controller of the world's oil supplies has upset the by no means smooth operation of the international monetary and trading institutions set up at the Bretton Woods Conference over 30 years ago. Many of the oil producing states are unable to absorb increased imports of industrial goods because of their small populations. As a result, some have accumulated vast amounts of currency for which investment outlets have to be found. This build-up of indebtedness to the oil exporters represents a redistribution of wealth, in favour of the OPEC states.

The nations regroup

A significant effect of the increasing economic power of the oil exporting countries is that it has made it necessary to reclassify the nations of the world. Formerly, three broad groups sufficed: the industrialized countries of the West, the Communist states, and the developing countries known collectively as the Third World. Now this last group is itself divided into three parts consisting of the OPEC states which, apart from their oil resources are largely undeveloped, the commodity-rich developing countries able to earn foreign currency by the export of commodities and minerals, and the poorer developing countries, now known as the Fourth World, which have no export surpluses and find the increased price of imported oil an added constraint on development. This regrouping has affected the operation of the various UN agencies and led to the emergence of the OPEC states as the potential leaders of the developing world. Yet another grouping has appeared, this time among the oil producing states. The first group is made up of countries which are in need of high oil revenues to finance development plans and other high priority spending: this includes Algeria, Indonesia, Ecuador, Gabon and Nigeria. The second group contains the 'medium need' countries, which can spend some but not all of their oil revenue on current investment and consumption: this includes Iran, Iraq, and Venezuela. Third, there is the group of states which, because of small populations or for other reasons, are building up surpluses of currency from oil revenues over and above their requirements: these are Saudi Arabia, Kuwait, the United Arab Emirates, Libya and Qatar. The grouping of rich and poor consumers on the one hand and needy, affluent, and excessively rich oil producers on the other, underlines the complicated situation.

World energy reserves

So far we have been concerned with oil, the most widely used fuel, but the solution of national and international energy problems must take account of the other fossil fuels – coal and natural gas – as well as nuclear energy and a variety of new sources of energy in various stages of development. Conservation of fuel has now become a major pre-occupation of the industrialized and developing countries alike.

In general terms, total world resources of energy are virtually inexhaustible, but in many cases the resources are inaccessible or cannot be brought into use in the present state of technical knowledge. Easily accessible reserves which can be marketed at a price that would-be consumers can pay all have a life span which can be calculated in terms of current rates of consumption. While it is true that increases in the prices of particular fuels will bring into production reserves previously not profitable, the usable supply of fossil fuels is limited. More than any other factor of production, fuels illustrate the fundamental problem of economics, that resources are finite while human needs are capable of indefinite expansion.

In the nineteenth century the existence of large supplies of coal in western Europe and North America enabled new industries based on the steam engine to be established. The importance of coal lasted until after the Second World War. The Schuman Plan, on which the European Coal and Steel Community and later the EEC were based, was conceived as a means of preventing any future war between France and Germany by placing the control of their coal and steel industries under a High Authority. Since then coal has come down in world energy ratings and been supplanted by oil. After two frantic decades of oil-based economic growth the industrialized nations are again turning to coal to fill the energy gap foreseen as oil production goes into decline from around the mid 'eighties. The nineteenth-century situation in which the possession of the leading fuel automatically gave industrial leadership has not been repeated for a number of reasons. Most of the major oil producing countries, except the US, have no industrial structure and were poor developing countries until the discovery of oil raised them to affluence. Oil is much more easily transported than coal so that it has become a major commodity in world trade. This trade has been carried on by a small number of multi-national corporations with production, refining, storage and distribution facilities throughout the world. The concentration of the world's major reserves of oil in a relatively small area in the Middle East has made it possible for oil pricing policies and supplies to be determined by political rather than commercial considerations.

In the energy world the role of consumer preference is constrained by decisions on the timing of investment in productive capacity and by political decisions on the level of production. In other words, while it is broadly true to say that energy will not be produced that cannot be used, there is no doubt at all that energy which for any reason is not produced cannot be used.*

Law of increasing intervention

The importance of securing adequate fuel supplies at commercial prices has meant that governments adopt energy policies designed to meet their particular requirements. Such policies involve state intervention to protect or stimulate supplies of indigenous fuel, action in anticipation of the possibility of interruption of supplies, and measures to control the use and price of individual fuels. There is no such thing as a free market for fuels and policy choices are concerned with the different types of government intervention that will be employed. In recent years the British government has become increasingly concerned with participation in production of North Sea oil, the control of investment by the nationalized coal, gas and electricity industries, and development of the nuclear power industries. The impact of intervention will be discussed later in this book in relation to individual fuels and to the energy situation over the next 25 years. The obligations of British membership of the EEC in connection with energy policy must also be taken into account.

Whether, as has been suggested,† a law of increasing intervention operates in relation to energy policy, there is no doubt that government control over the use of different fuels, their price, source, ownership and availability will continue to grow in the present situation. While a return to *laissez faire* would be impossible, even if it were desirable, it is important that intervention should, as far as possible, be confined to

* Workshop on Alternative Energy Strategies, Energy, Global Prospects 1985–2000, McGraw-Hill, 1977.

† Colin Robinson, 'Competition for fuel', IEA Occasional Paper 31.

problems in the fuel market which require government control, and that thought should be given to the forms of intervention most likely to produce the required results.

It is tempting to believe that, in a world where producers, whether of fuels or commodities, are increasingly able to secure ever higher prices from importers, state intervention in economic affairs is the natural order of things. This belief mixes up questions of government, which is about problem-solving, with politics, which is about issues. In the last 20 years every British government has introduced measures to control the supply and use of fuels. Looking back at what has been done to the coal industry, it is hard to believe that the measures taken have been particularly wise or far-sighted. However, in the absence of any means of knowing what would have happened without intervention, the choice lies between ends rather than means.

Looking to 2000 AD

The content of energy policy over the next 25 years has already been determined by events. The demonstration of monopoly power by OPEC in October 1973 serves, in energy terms, as a dividing line between two different epochs. In the previous two decades it had been possible to believe that increasing quantities of cheap oil would be available through the nineteen-eighties and beyond. The events since October 1973 have made it clear that the OPEC states are capable of acting together, and that so long as they do so they can exert considerable pressure on oil importing countries.

The problem facing the industrialized countries is how to manage the transition from dependence on oil to a greater reliance on other fossil fuels, nuclear energy and, at some later stage, renewable energy systems. The possession of North Sea oil alters the timing of this operation for Britain, but does not remove the necessity for action.

The meaning of the hidden agenda

From this summary of the diverse issues raised by the energy problem the meaning of the hidden agenda becomes clear. The formulation of an energy policy is no longer a matter of enunciating a principle, for example the 'four-fuel policy' of the 1967 White Paper. Policy repercussions now raise many very diverse issues: technological considerations of what is possible; economic and commercial priorities; welfare aspects of supply and use of fuels; pollution of the atmosphere; destruction of the environment; the political revolution determining who shall control energy policy; the wider international aspects of relationships within the OECD and the EEC; the North–South dialogue. All this is happening within the context of the time scale of known resources and the prospect of their exhaustion in a foreseeable future. It is not necessary to be a paid-up member of the Club of Rome to realize that the hidden agenda behind the facade of the preparation of a national energy policy is no more and no less than a checklist for our survival. Furthermore, the time during which adjustments can be made is running out. At some point within the next half century the world production of oil and natural gas will fall below the level at which total demand can be satisfied. Other fuels – coal, nuclear power, alternative energy sources – must be brought to the point where supplies are adequate. At the same time, international institutions will have to adjust to a new pattern of relationships between industrialized and developing countries. At the national level, economic and social priorities must be sorted out and decisions taken on the objectives of economic policy. The crowded quarter century that lies ahead is all the time we have to begin tackling the diverse items on the hidden agenda.

Chapter 2
British energy policy

Energy equals coal

Before the Second World War, Britain did not have or require an energy policy. Abundant coal was available at low cost to fuel the steam engines, locomotives and ships' boilers on which the economy depended, as well as to keep a fire burning in every hearth. Coal was energy in terms of the British economy. True, there were some inroads into the traditional monopoly of coal, notably the Navy's conversion to oil at the insistence of Winston Churchill, back in 1911. The nearest approach to an energy ministry was the Ministry of Mines with general responsibility for coal, and this did not merit a seat in the Cabinet. The coal industry was regarded as one of the 'commanding heights' of the economy when it was nationalized on 1 January 1947. The belief that control of the coal and steel industries of Europe was a determining factor in the preservation of peace was behind the Schuman Plan of 1950, from which the EEC subsequently evolved. The need for a British energy policy only arose as changes in industrial technology and the using up of the more readily available and therefore cheaper supplies of coal brought increased competition from oil and, later, natural gas.

The National Plan

The first detailed official forecasts of energy requirements five to six years ahead were published in 1965. Forecasts for coal production had appeared from time to time, and the building of the first nuclear power stations had followed the production of very broad statements of long-term energy demands. However, it was not until the preparation of the National Plan* and its publication in 1965 that any comprehensive forecasts were made linking energy demand to the growth of the economy. The forecasts were made by the Ministry of Fuel and Power with the help of the different fuel industries. They were then taken on by the Department of Economic Affairs which had been created by the Labour Government of 1964 under Mr (now Lord) George Brown. The National Plan published separate forecasts for coal, oil, natural gas, and nuclear energy and related these to the main sectors of consumption within the context of an increase of 25 per cent in gross domestic product from 1964 to 1970.

As an attempt to analyse the allocation of resources necessary to support a policy of faster economic growth, the National Plan was a remarkable document. Unfortunately it appeared at a time when the force of economic circumstances required rigid deflationary policies. It was not possible to expand our way out of balance of payments difficulties, support sterling, and generate the investment needed for faster growth all at the same time. This was recognized in the brief but revealing statement from the National Economic Development Council (NEDC) printed at the beginning of the National Plan. In the event, the Plan was virtually abandoned in July 1966 when the government had to take emergency measures to ease the overseas payments position. The Department of Economic Affairs had lost its brief battle with the Treasury. George Brown moved to the Foreign Office, and the Department slowly faded away.

National Plan forecasts

So far as energy was concerned, the National Plan forecast demand for 1970 both for energy as a whole and for each fuel. For the record, these can be seen in Table 2.1. The demand for coal, excluding exports, was expected to fall to 175 million tons a year while oil and nuclear energy both showed significant increases. In general terms the Plan assumed that the 25 per cent growth objective could be achieved with an increase of only 13½ per cent in energy consumption. This meant that 2·1 per cent more energy would be consumed each year between 1964 and 1970

* 'The National Plan', Cmnd 2764, HMSO, London, September 1965.

Table 2.1

Industrial Inquiry.
Industries' estimates of inland fuel demand in the UK

Million tons of coal equivalent*

	1960	1964	1970	Actual
Coal:				
for power stations	51·9	68·0	84·0	
for gasworks	22·6	20·5	10·0	
for other purposes	122·2	98·7	81·0	
Total	196·7	187·2	175·0†	154·4
Oil (including petroleum gases):				
for power stations	9·2	9·7	14·0	
for gasworks	1·9	5·0	14·5	
for other purposes	54·4	78·6	115·5	
Total	65·5	93·3	144·0	145·6
Natural gas	0·1	0·3	1·5	17·6
Nuclear and hydro-electricity	2·6	5·1	16·5	12·0
Total inland demand for energy	264·9	285·9	337·0	329·6
Gas (million therms)	2,636	3,014	4,635	
Electricity (thousand million kilowatt hours)	104·9	143·4	241	

* Except for gas and electricity.

† It is estimated that a further 5 million tons may, on present prospects, be exported.

Source: 'The National Plan', p. 120.

to support an economic growth rate of 3·8 per cent. The 'energy/ GDP coefficient' produced an increase of 0·55 per cent in energy consumption for every 1 per cent of economic growth (see the National Plan, page 120, para. 8).

In view of the well-founded distrust of energy forecasts it is interesting to note that the National Plan global forecast for Britain's total energy requirement was reasonably accurate, but its estimates for individual fuels were way out.

The discrepancies between actual and forecast production and use of coal for 1970 illustrate the problems of forecasting. Demand for coal, according to the NCB,* actually exceeded production, which was 5 million tonnes below the previous year and 16 million tonnes less than the National Plan forecast for 1970. The most important discrepancy was the fact that gas works took only 3·5 million tonnes instead of the 10 million forecast in the Plan. By 1970 the build-up in the use of natural gas accounted for 70 per cent of total gas consumed, compared with only 45 per cent a year earlier.

The 1965 White Paper

Although the National Plan was removed from the centre of the political stage, the attempt to formulate a comprehensive energy policy continued. Two fuel policy White Papers followed each other in quick succession in 1965 and 1967. Between them they effectively brought to an end the coal-based energy policies which had persisted into the 'sixties. The 1965 White Paper† presented a rationalization of policies for the principal fuels, which echoed the reservations about the future of coal contained in the National Plan. It stated that the development of 'a coherent fuel policy' was necessary because of technological advances and went on to spell these out:

> New raw materials and processes are transforming the nature and prospects of the gas industry; the generation of electricity by nuclear power on competitive terms is now within reach; a large expansion of oil-refining capacity is in progress; the coal industry, though it has notable technological achievements to its credit, faces difficulties in covering its costs and holding its markets.

The 1965 policy statement foreshadowed a number of problems that were to develop during the following decade, while looking back obstinately to the era dominated by coal as the 'normal' British situation.

* National Coal Board, Report and Accounts 1970–71, Vol. 1, HMSO, London, July 1971.

† 'Fuel policy', Cmnd 2798, HMSO, London, October 1965.

The coal industry

To understand the significance of the change of direction taken in the White Paper it is necessary to look back over the previous decade. In the 'fifties the British fuel economy was overwhelmingly dependent on coal. Of the relatively small amount of oil then used, over half was for transport, and the national requirements were met largely by the import of products refined abroad. Electricity and gas depended entirely on coal and the railways were still in the age of steam. The main concern was still to achieve the level of output for coal which had been put forward by the National Coal Board in its *Plan for Coal* in 1950 and accepted as the basis on which the industry would develop. The 1950 *Plan for Coal* forecast a demand for 240 million tons of coal a year in the period 1961–65, including 25–35 million tons for export. This estimate was declared too low by the Ridley Committee* which reported in 1952. The NCB accordingly revised its production possibilities and stated that it might be able to increase output to about 250 million tons a year for the period 1965–70.

In fact, the decline in coal production had already set in, and far from building up its level of output, the industry had difficulty in meeting demand both at home and for export. As a result, coal imports, which reached 11·5 million tons in 1955, had to be arranged at high cost, mainly from the US. Because of the industry's inability to meet demand, the use of oil in power stations was encouraged by the government, so that by 1959 they were burning over 7 million tons of oil a year, compared with under 1 million in 1956. Oil consumption generally increased in the mid 'fifties, a period of minimal economic growth. Inevitably this meant that oil was taking a bigger share of a static market, with a consequent sharp decline in coal sales accompanied by rising pithead stocks. Licences for the import of US coal were stopped, and the electricity industry was asked to halt the switch-over from coal to oil. The 1961 Budget imposed a duty of 2 old pence a gallon (about £2 a ton) on oil used for burning. This duty, conceived as a source of revenue, was retained for the quite different purpose of protecting the coal industry. It is still being levied at the time of writing. Thanks to these measures and the fact that the economy moved into a period of growth at the end of the 'fifties (Harold Macmillan's 'never

* Report of the Committee on the Use of National Fuel and Power Resources, Cmnd 8647, HMSO, London, 1952.

had it so good' years), the position of coal improved. Output rose to around 200 million tons a year and remained there for the four years to 1963.

The oil industry

From 1964 onwards the rise in oil imports and the fall in coal output proceeded side by side. The Central Electricity Generating Board (CEGB) and the Gas Council were again urged to take more coal, and a scheme was introduced to promote the use of coal for heating public buildings. From a peak of 217·5 million tons in 1956 coal production fell to 187·2 million tons in 1964. In the same period oil consumption rose from 37·5 million tons coal equivalent (c.e.) to 93·3 million tons c.e. in 1964. Only governmental pressure on the CEGB prevented an even greater decline in the power station coal 'take'.

Over the same period changes were taking place within the oil industry. In 1950 less than 10 million tons of crude oil were refined in Britain. This picture changed rapidly as new refineries were built, beginning with Fawley on the Solent. In 1956 refining capacity stood at 29 million tons, rose to 39 million tons in 1959, 53 million tons in 1963 and 58 million tons in 1964. This shift to importing crude oil rather than refined products reflected changes in the development of the international oil industry. The US, which had been the world's major producer of crude, and exporter of refined products, was beginning to concentrate on supplying its domestic market, while the major oil corporations were depending increasingly for supplies on their concessions in the Middle East. The price of oil was moving downwards and the general tendency in the years 1950 to 1964 was in favour of oil as against coal. This trend continued in spite of the imposition of the oil duty in 1961. The 1965 White Paper recognized that the increased dependence on oil carried with it risks to security of supply and increasing balance of payments costs. But the growth in the use of oil was accepted as being in the national interest, while measures were to be taken 'to mitigate the security and balance of payments disadvantages by ensuring adequate stocks in this country and by encouraging the oil companies to diversify their sources of supply and to develop the United Kingdom refining industry'. (Cmnd 2798, para 66.) The search for gas and oil deposits in the North Sea was noted in passing as a development to be encouraged.

The 1967 White Paper

The 1965 White Paper did little more than review the position of the various fuels and set out broad policy objectives. The 1967 White Paper* was a very different exercise. Its importance lies in the fact that it broke away from the coal-versus-oil argument and moved on to a four-fuel economy in which natural gas and nuclear energy had an important role. The White Paper acquired an importance of a different kind in that it was the last official statement of the fuel position to be published for a decade.

The 1967 White Paper appeared after North Sea gas had been discovered but before it had begun to come ashore in quantity. North Sea oil was still a remote possibility and the White Paper therefore concentrated on supplies of fuels actually being produced, or for which investment had been undertaken. Each of the four fuels covered by the White Paper had undergone important changes in the 'sixties. The National Coal Board had embarked on a programme of reorganization with the concentration of production on the most efficient collieries. At the same time new techniques for winning coal were being rapidly introduced, with increased mechanization at the coal face and power loading bringing increased productivity. Worked-out and uneconomic pits were closed at the rate of about 40 a year, with the majority of closures in the older coalfields in Scotland, south Wales and the north east and north west of England. The most serious aspect of the coal situation was the fall in the numbers employed throughout the 'sixties. At first this was achieved through natural wastage, but as time went on, with the labour force falling by some 25 000 a year, it became increasingly difficult to recruit labour for the mines and maintain a balance between age groups and skills. In the east Midlands, where well-paid jobs were to be had in other industries, the shortage was especially acute.

The development of the oil industry had reached a point in 1965 where it provided 37 per cent of Britain's total energy, compared with 15 per cent a decade earlier. In the early 'sixties there was a growing consumption of naphtha as feedstock for the gas and chemical industries, in place of coal. The arrival of natural gas checked this particular development. The greatest expansion was in fuel oil which had risen until it accounted for about a half of the total energy. About 40 per cent

* 'Fuel policy', Cmnd 3438, HMSO, London, November 1967.

of the oil imported was used for road and air transport where it has no substitute. In the decade to 1965 there was a substantial decline, in real terms, in the pre-tax price of fuel oil. The policy of encouraging home refining had been followed largely for balance of payments reasons, as it meant that the cost of refining was incurred in this country instead of abroad. It also gave increased security of supply, as it is easier to switch sources of crude than of refined products. There was the added advantage that it took account of the fact that much larger, and therefore cheaper, tankers could be used for transporting crude. The objective of government policy was that home refining should be on a sufficient scale to cover inland demand and bunkers, with imports of particular refined products balanced by exports of others. However, this favourable state of affairs was never reached, for a variety of reasons, especially commissioning delays.

Security of supplies

The Arab–Israeli war of 1967 and the outbreak of the civil war in Nigeria had raised the long-standing problem of security of supplies, while the White Paper was being prepared. Some doubts were therefore expressed both about availability of current supplies and the long-term adequacy of world oil resources. In view of what was to come in the decade ahead, the following example of official thinking on the subject from the 1967 White Paper, has a certain poignancy, which merits quotation at some length.

> There is certainly oil in the ground to meet the world's demand well beyond the period under review, and the danger is rather one of being denied normal supplies by political or other events outside the control of the industry or the Government. In the longer term this danger is limited by the fact that producing countries are at least as dependent on trade in oil as we are ourselves. But war or civil strife may cause interruptions and production or exports may be stopped for political reasons or in the attempt to gain a sharp increase in unit revenue, even though this might affect the long-term growth of outlets for oil. Statistical exercises are undertaken regularly to test the country's ability to survive interruptions of varying severity and duration. (Cmnd 3438, para 47.)

The reference to security of supplies ended with an expression of confidence in the ability of the oil companies to reorganize supplies and pointed out that the appearance of new sources, including Libya, had lessened dependence on the Middle East.

Few people reading the White Paper when it appeared paid any attention to this appraisal of the security question, or indeed saw it as anything but an example of careful official drafting on an aspect of energy policy that had to be noted on grounds of comprehensiveness.

Natural gas

Government policy on natural gas was that it should be brought into use as rapidly as possible. The responsibility for bringing the gas into use had been placed on the Gas Council and Area Gas Boards by the Continental Shelf Act of 1964. The government's view was that the price negotiated between the Gas Council and the producers should be fixed as low as possible consistent with encouraging exploration. The major financial problem involved was the need to convert the appliances of the 14 million gas consumers to enable them to use natural gas. The capital cost of this operation was estimated at £400 million over a period of 10 years. However, the use of natural gas to produce town gas by substituting it for oil products to enrich the lower calorific-value gas produced in the oil reforming process would have involved building new plants at greater cost than a programme of phased conversion.

Market surveys carried out by the Ministry of Power showed that the saving in resources from using natural gas instead of other fuels was greatest where it commanded the greatest price premium. This suggested that resource savings would be highest where natural gas replaced manufactured gas, intermediate where it had a price advantage over oil, and lowest in the bulk industrial market. However, the price of supplies to the gas industry depended on a rapid build-up of supplies and a relatively short depletion period. For this condition to be fulfilled, some gas would have to go to the bulk industrial market with consequent lower saving of resources. It followed that the rate of absorption of gas, the rate of depletion of the gas fields, the rate of build-up and the load factor at which supplies could be accepted were all closely linked with the price at which natural gas was bought by the Gas Council. The long-term benefits to the economy (in terms of foreign exchange savings, mainly on imports of oil) had to be set against the short- and medium-term costs of conversion of consumers' appliances and of the construction of a nationwide transmission system. In the context of the four-fuel policy of 1967, natural gas was still an untried factor, but one expected to increase security of supplies and reduce the quantities of oil imported. For the consumers there was no certainty that the cost advantages of natural gas would be reflected in lower prices.

In the decade since the publication of the 1967 White Paper, natural gas has become a major source of energy in Western Europe, where it had previously, in contrast with the US, not been in plentiful supply. The calculation of ultimate resources of natural gas is more difficult than for other fuels. Natural gas occurs either associated with oil or coal deposits, or by itself as well gas, as in the southern basin of the North Sea. The amount of gas associated with oil varies widely between oilfields. Furthermore, even where associated gas occurs, the geographical position of the oilfield may be such that it is far too expensive to collect the gas and move it by pipeline to terminals in areas of demand. It may be even more expensive to set up liquefaction plants to enable the gas to be transported in refrigerated vessels to overseas markets. It follows, therefore, that in present circumstances only a proportion, possibly less than half, of the associated gas in the world's oilfields can be regarded as a part of world energy reserves. The major gas reserves are in Russia, China, Eastern Europe and the Middle East. Reserves in the US are now not sufficient to meet demand and various plans are being implemented to import natural gas from the USSR and the Middle East.

Nuclear energy

It was on the demand for electricity and prospects for nuclear power that the forecasts of the 1967 White Paper strayed farthest from the real world. The maximum demand for electricity forecast for Great Britain for 1972–73 was set at 61 000 megawatts (MW) compared with the existing demand of 40 000 MW. In the event, the demand for electricity has risen more slowly and in 1976 had reached 42 000 MW.*

Of the CEGB's 43 000 MW of generating capacity in 1967, 31 700 MW was coal fired, 5 300 MW oil and dual fired, 2 800 MW nuclear,

* Corporate Plan 1976, CEGB, June 1976.

and the rest hydro-electric and other schemes. The first nuclear programme, which started in the immediate post-war years with a total of eight Magnox gas cooled reactors built at Calder Hall, Chapelcross and elsewhere, was a considerable engineering and technological success. Its contribution to the capacity of the electricity supply industry had, however, been marginal. The early stations had an output of 50 MWe and could not compete on production costs with new, conventional, coal fired stations, while their construction costs had been considerably higher. Even so, by 1964 the nuclear contribution to the British electricity supply was greater than that of any other country, including the US. The second nuclear programme launched in May 1965, was based on the Advanced Gas Cooled Reactor (AGR), and construction began on the Dungeness 'B' station. The neighbouring Dungeness 'A' was an existing Magnox power station. The CEGB had great hopes for the AGR stations which were expected to produce electricity 'more cheaply than even the most favourably-sited contemporary coalfired stations'.* The Ministry of Power decided that one nuclear power station a year would be commissioned from 1970 to 1975, which would have brought the total capacity for the second nuclear programme up to 8 000 MW by 1975.

In a comparison between conventional and nuclear power stations, the 1967 White Paper claimed that total generating costs at each of the first two AGR stations would be less than those of the coal fired station at Drax on the Yorkshire coalfield, and the oil fired station at Pembroke, both then under construction. The comparison is shown in Table 2.2. Other reductions were expected for the period after 1975 from further developments in the AGR system. The White Paper also reported that a commercial scale (250 MW) fast reactor at Dounreay would be completed in the early 'seventies. The choice between nuclear and conventional power stations must depend on the movement of fuel costs and the development of nuclear technology. Construction costs for the later AGR stations were forecast to be little more than those of conventional stations, while their running costs would be much lower. As things turned out, only one statement on the second nuclear programme in the White Paper was wholly born out in practice. This said that 'no commercial scale AGR is yet completed and operating, and in a young technology the risk of disappointment must exist' (Cmnd 3438,

* Cmnd 3438, page 16, para 31.

Table 2.2

Generating costs of nuclear and conventional power stations

	1967 pence per kWh
Nuclear	
Dungeness 'B'	0·52
Hinkley Point 'B'	0·48
Coal	
Cottam	0·53
Drax	0·56
Pembroke (with tax)	0·53
Pembroke (without tax)	0·42

Source: 'Fuel policy', Cmnd 3438, Appendix III, HMSO, London, 1967.

page 17, para 33). All five AGR stations ordered ran into serious technical and other difficulties. The first AGR stations actually to be commissioned was Hinkley Point 'B' in April 1976. Work on Dungeness 'B' continues at the time of writing. As a result of the AGR programme, Britain lost her lead in the development of nuclear power stations. The British nuclear industry had to be reorganized and its original three separate groups were replaced by the National Nuclear Corporation. As will be seen later in this book, the AGR reactors have been superseded by the boiling water and steam cooled systems, both of which were developed outside Britain.

Uranium reserves

Consideration of the fourth fuel – nuclear energy – raises the question of the availability of uranium, the main source of fuel for the present installed fission reactors. World uranium reserves have been estimated at various cost levels. At a cost of up to $26 per kg of uranium, the World Energy Conference held in 1974 published a survey setting the total supply at 0·98 million tonnes. The uranium ores at lower grades than those giving uranium at $26 per kg could yield another million tonnes of uranium at prices around $2.39 per kg. If very low grades of uranium ore could be marketed, then it is possible that reserves could run to millions of tonnes of uranium. The working of very low-grade ores, however, could require the use of so much energy that extracting the uranium would not produce a net addition to world energy resources. The development of fast breeder reactors which produce more fuel than they burn is regarded as the best solution to the uranium supply problem, provided that such reactors can be safely constructed and operated.

The four-fuel policy

The four-fuel policy set out in the 1967 White Paper required that the fullest advantage should be taken of the new primary sources of energy, natural gas and nuclear energy. Coal consumption would be allowed to fall as costs increased, while the availability of natural gas and nuclear energy would ensure that Britain was not too dependent on imported oil. The rate of contraction of the coal industry was forecast at about 35 000 men a year. Demand for coal by 1970 was estimated at 146 million tonnes, rising to a possible 155 million tonnes as a result of increased productivity. Some 3 to 6 million tonnes would go for export. Electricity generation was seen as 'the only sector where increased coal consumption might occur'. The cost of additional coal for the electricity industry, over and above what would be used on a straight commercial basis, would be met by the government, up to a total of £45 million.

The 1967 paper on fuel policy was an honest attempt to take account of the changes taking place in the supply of primary fuels. For the first time the weakness of the competitive position of coal was accepted as a reason not for increased protection for the industry, but for slimming it down to a size where high productivity and low costs enabled it to make a profit. Where the policy makers went wrong was in believing that the contribution of natural gas and nuclear energy would quickly increase to a point where the dependence on imported oil would be reduced to manageable levels. As will be seen, the fact that no further policy statements were issued for the best part of a decade was less a tribute to the soundness of the 1967 White Paper than an indication of the difficulty of framing policies against a background of rapidly changing conditions.

Energy conservation

Up to 1973 the question of conservation of energy was not a serious element in policy making. Inevitably, with oil and coal not only readily available but cheap, there was no compulsion to economize in the use of fuels. This was true of research into the efficient use of energy, with one or two exceptions, such as the NCB's pioneer work on fluidized bed combustion. Projects featuring hydro-electric power development as part of a wider scheme for land reclamation and drainage were discussed in relation to Morecambe Bay and the Wash, as well as schemes for Severn and Thames barrages. None of these got very far before 1973 because of unfavourable cost-benefit situations, and since then inflation has priced them out of the market. District heating schemes, which remained at the pilot project stage, are another example of energy that was wasted because cheap supplies were available. The insulation of houses was undertaken as a means of avoiding burst pipes in the winter time, rather than as a positive step towards fuel economy. If full use is made of these and other positive methods of fuel conservation instead of relying on the exhortation to 'save it', we may arrive at a position in which energy is used sensibly instead of the all-or-nothing attitude exemplified by the finger on the switch. The fact that conservation was not taken seriously up to 1973 will make it all the harder to adapt to less profligate attitudes in the future.

Chapter 3
The hostile North Sea

The Continental Shelf

In the school geography books the North Sea used to be the place the trawlers went out to from Grimsby and Hull to catch fish. The seaside resorts on its shores are 'bracing' rather than relaxing, and all have an aura of purposeful, earnest endeavour. It was therefore fitting that when exploration for natural gas and oil began on the European Continental Shelf, the major discoveries were made below the cold waters of the North Sea, bringing new activities to areas that had been missed out of earlier phases of industrial development. As the oil bearing formations were at depths and distances from the nearest coastline not previously regarded as feasible, new technologies had to be devised to drill the oil wells. A whole new industry was created to manufacture the rigs and platforms needed to bring the oil to the surface, with development for the most part based on US technology and personnel.

Exploration of the North Sea using seismic techniques began in the early 'sixties. The area was divided up into blocks and exploration and production licences granted. Those involved were the major oil companies plus a number of consortia in which independent oil companies joined with merchant banks, industrial and commercial firms, and public sector bodies, including the National Coal Board. The discovery of gas, not oil, came from 1965 onwards in the southern basin of the North Sea.

North Sea gas

The discovery of North Sea gas did not come about by accident. Esso and Shell had been drilling in Holland for a number of years when they found the vast natural gas field at Schlocteren in the Groningen province of North Holland. Drilling in Britain revealed small quantities of gas at Whitby on the Yorkshire coast. This discovery led to the conclusion that the gas bearing formations might extend from coast to coast under the Continental Shelf. Up to that time the North Sea had not been regarded as a likely area for gas and oil prospecting. It was stormy and windswept for eight months of the year and its waters were deeper than those of other continental shelves already prospected. However, further exploration revealed promising structures on the sea bed and prospecting began in earnest. The first discovery on a commercial scale was made by BP in December 1965. The gas found in the southern basin of the North Sea is well gas, found without oil, trapped in reservoirs formed in porous limestone or sandstone under some impervious rock formation. The other kind of natural gas discovered in the North Sea is found in association with oil and has to be removed or flared off before the oil can be used.

The countries surrounding the North Sea agreed under the Geneva Convention on the boundaries for their own sectors, in which they each have control over exploration and production of oil and gas. The Norwegian and British sectors are the only ones to have reached large scale production, following heavy investment in capital equipment. The natural gas consists in the main of methane, an inflammable gas composed of four atoms of hydrogen combined with one of carbon. The gas is similar to the methane known as firedamp, formed in coal mines.

The Dutch example

Compared with town gas, natural gas has a number of important economic characteristics. First, the gas is found in a natural state and does not require any manufacturing process. Second, as natural gas is found at extremely high pressures, very little labour is needed to bring it to the surface. Unlike coal mining, where half the cost of making the fuel available is covered by the labour of cutting the coal and bringing it to the surface, the labour cost of producing natural gas is relatively small. However, the equipment needed for the drilling of the wells, first in exploration and then in developing a successful find, including the pipeline to bring the gas ashore, and the terminal and compressor stations joining it to the onshore distribution system, are all highly expensive items. The cost structure of natural gas is therefore quite different from that of town gas made from coal or oil based chemical feedstock. At the same time, natural gas competes directly with oil and coal as a source of heat and power. A company producing both oil and gas is therefore faced with a conflict of interest. In the case of the British sector of the North Sea, the producing companies were in a special situation as they had to negotiate a price for all supplies landed with the Gas Corporation,* formerly the Gas Council. They were not free to export the gas to other countries bordering the North Sea. In Holland, Shell and Exxon joined with the government, represented by the Dutch State Mines, in a partnership covering production and distribution. The Dutch government was particularly anxious to maintain control of the rate of depletion so that the advantages of possessing a major natural resource were not dissipated too quickly. The gas is sold to a distribution agency called Gasunie, which is owned 50 per cent by the companies, 40 per cent by the Dutch State Mines and 10 per cent by the government. Gasunie supplies the municipalities and large industrial users. The price of gas has been fixed at a level high enough not to upset the market for oil and coal. Exports to Belgium and West Germany are also handled by Gasunie.

Controlling UK supplies

This kind of joint venture arrangement is quite different from the system introduced in Britain. Under the Continental Shelf Act of 1964, the state owns the right to all minerals under the sea – except coal, the rights in which are vested in the National Coal Board – and has the sole right to award licences for exploration and development. Production licences are granted to companies on the basis of exclusive rights in the area covered for six years. After that, half the original area of the concession must be returned to the state, while the rights of exclusive exploitation of the remainder may be renewed for a further 40 years. All gas has to be sold to the Gas Corporation, which is responsible for its distribution. The only exception to this procedure is the sale of gas for

* Under the Gas Act of 1972 the British Gas Corporation replaced the Gas Council and the twelve Area Boards on 1 January 1973.

use as chemical feedstock. The Gas Corporation is responsible for all downstream investment from the point at which the gas comes ashore. The high pressure gas transmission system was constructed mainly after 1960, when oil gasification began to replace carbonization as the accepted method of producing gas. The grid pipelines are 36 inches in diameter but later construction will use pipes up to 48 inches. The gas distribution system built up over a period of 150 years, now extends to some 120 000 miles.

In the negotiations over the price of North Sea gas, the producing companies and the Gas Corporation started with very different viewpoints. The companies applied to gas the same broad formula which they used in negotiating oil prices. Oil companies make high profits from production, and use these to finance refining, distribution and transport. Profits on successful wells are expected to cover losses on unsuccessful ones. The Gas Corporation, on the other hand, wanted to concentrate profits on the distribution rather than production of gas, in order to finance the cost of converting existing appliances to use natural gas, and installing the new gas distribution system. The Area Boards paid the same price for natural gas regardless of their distance from the onshore terminal, and the quantities involved. The only price difference was created by the load factor, whereby those boards able to keep their pipelines operating at near full capacity paid less than those unable to do so. The difference in outlook was reflected in the starting prices when the negotiations began. The companies were thinking of over 4 (old) pence a therm, the Gas Corporation of only 2 (old) pence.

The much higher prices negotiated in 1973 for gas from the Frigg Field reflected the greater costs involved for associated gas.

Prospects for North Sea gas

As soon as it was clear that considerable supplies of natural gas would become available, the Gas Corporation had to come to a decision on how it should be used. Two broad possibilities existed, each with its own technical and economic advantages and disadvantages. The first was to use natural gas as a feedstock for the production of town gas. The second was to use the natural gas directly as a fuel and convert the whole distributive network and consumers' appliances for this purpose. The Gas Corporation chose the second of these alternatives, mainly because it made available a fuel with twice the calorific value of town gas at a cost well below that of efficient oil gasification plants. The higher calorific value of North Sea gas meant that the carrying capacity of the distributive pipeline network would be doubled. The rejection of town gas reflected the needs of all parties concerned. The producing companies required an early return on their capital, the Gas Corporation wanted to have a homogenous and cheap fuel to offer, and the government needed to reduce balance of payments costs by saving on naphtha and other oil based feedstocks used in the production of town gas.

The introduction of natural gas involved the conversion of appliances in the homes, offices and factories of some 14 million existing consumers, or, where this was not possible, the installation of new equipment. At the same time, terminals at points on the east coast where the gas was brought ashore had to be built, and large diameter pipelines laid to connect them with the main areas of population. Beginning in 1967 when the first gas was landed, the North Sea gas programme has been carried out well on schedule. By the end of 1975 it was supplying 4000 million cubic feet per day, in line with the estimate set by the 1967 White Paper. This was equivalent to almost 60 million tons of coal a day, four times as much gas as was being sold a decade earlier. A build-up on this scale was too rapid for all the gas sold to go to premium markets. Some non-premium sales were made, particularly to industrial users, on an interruptible basis. Some gas went to CEGB power stations. However, approximately three-quarters of sales went to premium markets, with over half the total of 13 000 million therms going to domestic users. Natural gas has increased its share of the total energy market in terms of useful heat for final users, excluding transport, from 8 per cent to 25 per cent. Gas is now the largest supplier of energy to the domestic sales market, with over 14 million customers. Proven reserves of natural gas are believed to be 28·7 trillion cubic feet, probable reserves 11·5 trillion cubic feet and possible reserves 10·3 trillion cubic feet, giving a total of 50·5 trillion cubic feet.*

Finance

By the end of 1975 it was true to say that North Sea gas was an outstanding success in all ways except one – it was losing money. The

* 'Development of the oil and gas resources of the UK' (the Brown Book), HMSO, London, 1976

contracts negotiated in the 'sixties had been overtaken by inflation, and the losses were not made good by the price increases of the mid 'seventies. In the financial year ending March 1974 the British Gas Corporation made a loss of £44 million. The Annual Report of the Corporation put the blame for this unfortunate state of affairs on the government's policy of price restraint at a time when inflation was boosting operating costs. Two price increases in 1975 went some way to restore the balance, although not before meeting considerable opposition from the National Gas Consumers' Council. While it may be argued that low gas prices can be a major force in holding down living costs, there seems little point in a situation where the more customers the industry manages to supply, the greater its losses. As an equally rigid price discipline has not been applied to coal, electricity and oil, a pricing policy which reflected the production costs of gas more accurately would seem to be to the competitive advantage of all four. Gas moved back into profit in 1975–76 (£25m) and announced a price increase of 10 per cent as its contribution to answering the criticism that gas was being sold too cheaply. Subsequent price increases have further reduced the price advantage of gas over coal and oil. Some idea of the change that has taken place in the gas industry can be seen from the fact that 10 years ago it needed 28 000 employees to produce 3·8 billion therms. In 1976 a natural gas supply of 11·4 billion therms was produced by under three thousand workers.

Expansion of gas industry

The cheapness of gas enabled the industry to expand into all domestic and premium industrial markets at a rate which has come near to overtaking available supplies. Since the beginning of 1973 the industry has been reorganized with the Council and Area Boards replaced by the British Gas Corporation. This change was intended, among other things, to give the industry greater commercial freedom. In theory the potential share of gas in the domestic market could be as high as 75 per cent, but this is likely to peak at about 60 per cent by the late 'eighties. The greater part of the increase in gas sales comes from domestic central heating systems.

Future supplies

The big doubt over the otherwise amazingly successful development of North Sea gas concerns the question of future supplies. Estimates of reserves have to be treated with caution and, paradoxically, it is only when they are exhausted that it will be possible to tell how big the reserves were. The level of gas supplies from the fields first exploited reached a plateau in 1975 and will decline slowly from 1980 onwards. This means that the Gas Corporation has had to plan ahead to bring new reserves into production. These will come from the Brent Field in the northern basin and the Frigg Field which overlaps the British and Norwegian sectors. The presence of the Norwegian trench to the east of the field prevents gas being piped from it to Norway. While it is anticipated that gas from these fields will be available, there is uncertainty about the amounts and timing. Whether much larger quantities of gas than are currently available could be disposed of would depend on pricing policies. Experience in the US and Holland would suggest that, given a reasonable price, gas can obtain a much greater share of the market if available in sufficient quantities.* If supplies are not available in the longer term, or if there is a delay in bringing supplies from new fields onto the market, it may be necessary to use substitute natural gas (SNG) or imported liquid natural gas (LNG) instead of North Sea gas. SNG is produced from coal or naphtha, and the Corporation is already importing LNG on long-term contract from Algeria. It would appear, therefore, that, even if North Sea gas supplies are interrupted or fall off for any reason, the Gas Corporation would be able to maintain its share of its premium gas markets.

Frigg and Brent Fields

Turning to the future, the major problem is the time of arrival of gas from the Frigg Field. This is 220 miles from shore and in 400 feet of water, so that heavy investment is needed to bring the gas ashore. As this is associated gas, the pattern of delivery is determined by the rate of extraction of the oil, rather than by the requirements of the gas market. Delays occurred in the early stages of the offshore work, and as Frigg is expected to increase supplies by about 40 per cent, these could have

* Sir Arthur Hetherington, 'The role of gas in Britain's energy market', *Coal and Energy Quarterly*, No.5, Summer 1975.

created serious difficulties. However, provided no more delays occur, the Frigg gas should ensure supplies into the 'eighties. It may be that some SNG will be required from time to time at peak periods, and greater demands will be made on the fields in the southern basin. SNG supplies could be made available by converting old town gas plant. Development of Frigg gas is due to begin in October 1977, with sales to industrial buyers promised for April 1978. On the marketing side, any hold-up in Frigg gas will mean that new sales to industry will have to be held back to allow the expansion of domestic sales to continue. The Gas Corporation has arranged a contract for the purchase of some gas from the Norwegian sector of the Frigg Field.

Following on behind the Frigg gas is the development of the Brent Field east of the Shetlands. This presents a similar set of problems, as the gas is found in association with oil. If the efforts to recover Brent gas go forward without any untoward happenings, attention will be given to the possibility of piping associated gas from other fields. One suggestion is for a large diameter 'gathering' pipe to collect the gas and take it ashore from a number of oil fields. The major uncertainty with these proposals, as with the development of the Frigg and Brent fields, is the timing and cost of bringing gas ashore. However, provided the estimated known reserves of 36 to 40 trillion cubic feet of gas actually prove to be available, North Sea gas should meet premium demand until the late 'eighties. With reasonable conservation measures in operation, and restriction of gas to premium uses, supplies could last well into the 'nineties. In any case, the gas industry is likely to be relying increasingly on SNG, and to a much less extent LNG, from the late 'eighties onwards. As North Sea supplies run out, the industry will be back to where it was before the first gas came ashore in 1967, with the difference that oil will no longer be available as a feedstock for gas making. Coal will be the obvious feedstock for SNG, but it will certainly not be cheap. Furthermore, the high capital cost of coal based plant will be an inducement to restrict the use of SNG to base load operation. One development that will take place over the next decade is the building up of a joint research programme on coal gasification between the NCB and the Gas Corporation. The *Plan for Coal* provides for the creation of new capacity with some of the additional supplies going to the gas industry for the manufacture of SNG.

North Sea oil

The search for gas and oil in the North Sea quickly established the fact that a major gas province existed in the southern basin, while oil as well as gas were to be found in the northern basin. The movement of the drilling rigs to the northern waters off the shores of Scotland and Norway took place from the late 'sixties onwards. The major discoveries of Ekofisk, Brent, Forties, Ninian and Stratfjord followed in rapid succession. The British sector, on the basis of existing licences, not only contains reserves estimated at 2·39 billion tonnes (17 billion barrels) of recoverable oil but some 40 trillion cubic feet of associated gas as well.* This great success story in exploration and discovery is matched by a less happy tale of misfortunes and mistakes in exploiting the reserves and bringing the oil ashore. A new hazard was brought abruptly to public attention by the blow-out at Ekofisk Bravo in the Norwegian sector in April 1977.

However, it is quite likely that additional reserves will continue to be established for some time to come, but at a less breathtaking pace. Large areas are still unexplored in the north as well as the seas west of the British Isles and in the Celtic Sea.

In terms of British energy resources the North Sea finds are considerable. In global terms they are less impressive. Britain has only about 2 per cent of world oil reserves compared with Saudi Arabia 25·2 per cent, Kuwait 12·2 per cent, Iran 11·4 per cent and the US 6·6 per cent. Estimates of the size of North Sea oil reserves vary according to forecasts of the rate of new discoveries. Government estimates put proven reserves at 1·35 billion tonnes, and 2·29 billion tonnes when discoveries with a better than 50 per cent chance of economic production are included. If North Sea areas not yet licenced are included, the total reserves could be over 4·5 billion tonnes.† This leaves the areas to be designated in the English Channel and the Celtic Sea to be accounted for.‡

* 'Development of the oil and gas resources in the UK', (the Brown Book), HMSO, London, 1976.
† The Brown Book, HMSO, London, 1976.
‡ The dates of the first five rounds of offshore licences were: First – September 1964; Second – November 1965; Third – June 1970; Fourth – December/ March 1972; Fifth – February 1977- .

However, there is a world of difference between oil that is flowing through the pipelines and oil that appears only in the far right-hand columns of statistical forecasts. Energy forecasts are notoriously fragile and it must be remembered that oil proved to exist may not be commercially exploitable at current prices. Even more oil would be left in the ground if oil prices were to fall as rapidly as they rose in the last quarter of 1973. Making allowances for rising costs of production, possible changes of government policy, unforeseen technical difficulties, programme slippages, and not forgetting the weather, the reserves in existing licenced areas should produce between 95 million and 115 million tonnes in 1980.* In 1975, the first year of significant production, 1·1 million tonnes of oil came ashore, mainly from the Forties Field. This figure was at the low end of the forecast in the 1975 Brown Book but, thanks to the prodigious development work done by BP, North Sea oil production was in the target area for the first time. In 1976 the five operating fields, Argyle, Auk, Montrose, Piper and Forties, were producing oil at the rate of 1 million tonnes a month.

By no stretch of the imagination can the North Sea be claimed as one of the greatest oil discoveries of all time. In 1973, at the time of the 'oil revolution', the existence of North Sea oil appeared as a singularly appropriate miracle designed to maintain Britain's claims to a position in the front rank of industrialized states. Here, in these unpromising waters, was a new indigenous resource which could make Britain self-sufficient in fuel, not to mention transforming Norway into a major oil exporter. Higher oil prices revolutionized the economics of offshore production, changing almost overnight the findings of the feasibility studies of even the relatively small fields. 1973 was the year when the possibility of a threat to the way of life of the industrialized countries was first glimpsed. In 1974 this danger appeared to recede so far as Britain was concerned, as discoveries of substantial new reserves in Ninian, Magnus, Stratfjord and the rest appeared to point the way to a more secure future.

Rising capital costs

The opening burst of enthusiasm for North Sea oil did not last much beyond mid 1974. Rapid inflation – at a higher rate in Britain than

* The Brown Book, HMSO, London, 1976.

elsewhere in Europe – and the unprecedented rise in capital costs for exploration and development removed some of the expected financial benefits to Britain of high oil prices. This was truer of the small fields, many of which are now only of marginal interest, than of the large ones already under development. The sharp lesson of the balance sheet led to a re-examination of the scale of North Sea reserves and likely production. It was established that sufficient recoverable reserves exist in the British sector to sustain a production rate of over two million barrels a day, enough to make Britain self-sufficient. The question that began to be asked was, when will the oil start coming ashore on this kind of scale? The forecast date moved back in a 'this year, next year, sometime' manner. Self-sufficiency, forecast for 1978, moved back nearer the 'eighties. Britain imports roughly 85 million tonnes of oil a year. Production of this amount from the North Sea would not mean that the market was completely supplied as it would still be necessary to import some heavier crude oils to mix with North Sea crudes in the refineries. Some North Sea oil will be exported as high quality light crude, most probably to the US and the EEC.

Taxation policy

A major complication was introduced by the government's policy with regard to taxation. The main feature of the system is the Petroleum Revenue Tax (PERT), designed to ensure that a sufficient proportion of the proceeds go 'to the British people as a whole', whatever that means. The monthly Economic Progress Report for May 1975, published by the Information Division of the Treasury, showed that the Government 'take' on a hypothetical North Sea oilfield would be 73 per cent, leaving 27 per cent for the companies. This would be made up of royalty at 12½ per cent of the well-head value of the oil, PERT at 45 per cent and corporation tax at 52 per cent on net revenue after deduction of royalty, PERT and expenses computed according to normal corporation tax rules. The report stated that while it is difficult to calculate the income of the government from the North Sea, as this will depend on the oil price level, costs and volume of production, it is likely to average some 70 per cent of oil companies' net revenues. Between 1975 and 1980 'the total Government 'take' might be some £3000 to £4000 million, of which PERT would amount to some £900 to £1300 million'. During the early 'eighties it might be about £2000 to

£3000 million a year, of which PERT might amount to £800 to £1200 million a year. Some relief is provided through marginal field safeguards intended to ensure that the rate of return on these fields is sufficient to encourage their full development. There is an oil allowance of one million tonnes a year for all fields which will be free of PERT subject to a cumulative total of 10 million tons per field. There is also a safeguard provision as a protection against a fall in the oil price or a further dramatic increase in costs. Events have proved that these forecasts are likely to be substantially correct. On 1 March 1977, the Energy Minister, Mr Anthony Wedgwood Benn, announced in the House of Commons that royalties from North Sea oil and gas yielded a total of £66·6 million in 1976, of which £44·2 million came from oil. This means that, with PERT and corporation tax, some 70 per cent of the oil companies' profits arrive in the Treasury.

BNOC

Under the Petroleum and Submarine Pipelines Act the government took powers to enable it to participate in North Sea oil development through the British National Oil Corporation (BNOC). This new body was to explore for, produce, transport and refine, store, distribute and buy petroleum; carry out consultancy, research and training; and build, hire or operate refineries, pipelines and tankers. The government also began negotiating with oil companies to obtain major state participation in existing oilfields. Where there is such participation, the government's 'take' includes proceeds from the sale of its oil.

The British National Oil Corporation was formally established on 1 January 1976. Its purpose was to hold the public stake in current and future licences acquired under the government's policy of participation in North Sea oil. It was also empowered under the terms of the Petroleum and Submarine Pipelines Act to extend its activities into refining, marketing and other operations. The government policy on participation is to secure a state holding of 51 per cent in existing oilfields by negotiation. An option to take a majority partnership is a condition of future licences for all commercial fields. If BNOC becomes a sole licensee in a particular field, all the profits of the licensee, and not just 70 per cent of them as elsewhere, will accrue to the state.

BNOC acquired sizeable assets on its formation by taking over the North Sea interests of the National Coal Board, giving it a stake in the

Thistle, Dunlin, Hutton and Stratfjord (extension) fields, as well as the gas producing Viking field. The NCB acquisition cost BNOC £50 000 which represented the nominal value of the equity capital of NCB (Exploration), and about £90 million to cover the outstanding loan obligations of NCB.

The great value of the NCB interest was that it provided BNOC, which then existed largely on paper, with an experienced staff with expertise in the oil industry, as well as acquiring the NCB's North Sea installations. BNOC also took over the 21 per cent interest of Burmah Oil in the Ninian field with estimated recoverable reserves of about 1000 million barrels. BNOC has moved into action only gradually since its formation was announced. The chairman and chief executive, Lord Kearton, was a former chairman of Courtaulds, and the first deputy chairman, Lord Balogh (who resigned in 1976 on grounds of age), is an economist who acted as adviser to Mr (now Sir Harold) Wilson in the 1964–70 Labour Government, and was for a time Minister of State at the Department of Energy when Labour returned to power in 1974.

Control of resources

The decision to set up BNOC was part of the policy commitment to bring ownership of North Sea oil under public control. It was the Labour Government's response to the need to deal with indigenous supplies of oil on a different basis from policies that had been adequate when all oil was imported under arrangements made by the international oil corporations. In the new situation the speed of development, the location of new refining capacity and the disposal of crude and products in the home or overseas markets were all subjects to be determined by the state. Of the four previous allocations of licences, those of 1964, 1970 and 1972 had been carried out by Conservative governments and only 1965 by Labour. The White Paper on UK Offshore Oil and Gas Policy (Cmnd 5696) July 1974 pointed out that foreign corporations had received most of the licences. There is a certain amount of hindsight about this argument, as when the earlier rounds took place it was not certain that oil would be found, or that it could be produced from such deep waters. There is a precedent and parallel for this action in the case of Norway which has set up a new state-owned company, Statoil, in preference to using Norsk-Hydro in which it had a 51 per cent holding. In Britain the claims of Scottish national groups to control over 'Scottish oil', meant that production and distribution could not be left entirely to the multi-national corporations.

Doubts about BNOC

While there were good reasons for the state taking some control over the development of oil and gas from the North Sea, doubts were expressed whether the creation of BNOC was the best way to do this. One main criticism of BNOC is that its creation involves drawing away resources from BP. The arrangement under which BP is to train personnel for BNOC came in for adverse comment on the grounds that BNOC had not been able to attract anybody of significance with oil industry experience to its own staff. A policy of introducing tighter state control over BP will adversely affect its ability to act as a multi-national oil corporation, which it has done most successfully in the past. One particular problem is the major shareholding in the Alaskan North Slope development held by BP. It is unlikely that the US government will accept a British nationalized industry, if this is what BP is eventually turned into, in that role.

Another set of difficulties arises over the financial activities of BNOC. As it takes up its 51 per cent participation agreements it will have to find considerable sums. A borrowing limit of £600 million was imposed on BNOC for its first years, but this can be raised to £900 million with the consent of Parliament. Even the higher figure is likely to be much too small to meet all the financial obligations which BNOC has taken on. The take-over of the NCB holding meant that BNOC had to assume some 27 per cent of the cost of developing the Murchison Field, calculated at around £500 million. As a 51 per cent partner in all new licences, BNOC has to pay its share of both exploration and development costs. To meet these charges BNOC can borrow from the National Loans Fund or on the commercial money markets. It can also draw on the National Oil Account, the fund into which the proceeds of royalties and licence rentals are paid. BNOC also has the benefit of its own revenues from oil produced in the fields in which it already participates.

Although BNOC has a choice of sources of funds its financial position will always be determined by the number of development projects in which it is involved. Further heavy expenditure will be required to meet the automatic provision contained in the terms for the fifth round of licence awards, that BNOC will provide 51 per cent of the cost of exploration in new areas. In the earlier North Sea concessions, dry holes were written off at the oil corporations' expense, and cost of successful holes was met from profits. After a quiet start, the BNOC operation is now beginning to take shape in terms of oil and finance. BNOC will apparently seek control of as much crude oil as possible. Its first Report and Accounts (May 1977), showed involvement in a wide range of interests including the North Sea operations of the NCB and Burmah Oil, participation agreements in all existing fields, a 51 per cent equity in fifth round licences, and production licences covering 59 blocks or part blocks. These acquisitions were paid for through the National Oil Account. In June 1977, BNOC reduced its dependence on public sector funds by raising an $825 million loan from a group of US and British banks. A US corporation, Britoil, registered as a charity for tax purposes, has been set up to handle forward purchases of oil to cover the loan. This operation enabled BNOC to repay its high interest public sector borrowings and at the same time take an important step towards establishing itself as a national oil corporation with international ambitions.

What happened to natural gas

The discovery in the 'sixties of natural gas was of immense significance for the British energy economy, resulting in a fourfold increase in gas supplied measured in therms. Gas now supplies 17 per cent of primary energy but, because of its relative efficiency, it accounts for about 25 per cent of useful heat consumption.

The economy can be said to have benefited from the switch to natural gas by import savings of about £2·3 billion a year on imported oil, representing 2·1 per cent of GNP. The main success of North Sea gas had been in taking over much of the space heating market from oil, coal and electricity. This is a new market, the extension of which is largely due to the availability of natural gas, with its considerable advantages for this particular operation. In premium uses in the industrial markets natural gas has been particularly successful in competition with coal, less so with oil and electricity. Gas for industry rose from just over one-fifth of total gas sales to almost a half between 1970 and 1976. One advantage of this activity is that it involves selling to only 70 000 customers as against 13·6 million domestic users.

The great success of North Sea gas raises questions central to the formulation of energy policy. Has the cheap gas which became available in the 'sixties been used to the best advantage? Did its existence and the low prices policy followed hold back other developments in the energy field? In particular, the availability of gas meant that progress in the coal industry was held back, and there was no incentive to conservation or to the development of alternative energy sources – winds, the tides, the sun, etc. There would also have been a greater incentive to get on with the development of the third generation of nuclear reactors, and so maintain Britain's leading position in the nuclear industry, if it had not appeared that we already had a new fuel of great efficiency in premium uses. In other words, energy policy is just as much concerned with what might have been, as with what is happening and is going to happen. Would it have been better for coal and electricity if a slower rate of depletion of natural gas had been enforced by charging higher prices, with, perhaps, a gas tax? Or would this have shifted the demand for bulk energy to imported oil, with a consequent increase in the balance of payments deficit?

Future for North Sea oil

While natural gas came ashore at a time when imported oil was plentiful and cheap, the arrival of North Sea oil has had a very different welcome. It is possible that, if the same compulsions had existed in the late 'sixties, the treatment of gas supplies would have been different. Gas was dealt with not as an essential element in a long-term energy policy, but as an important aid to the balance of payments and an additional factor in the drive for faster economic growth. North Sea oil, by contrast, came to us in the shocked aftermath of the OPEC price revolution. While gas was a commercially respectable addition to our national resources, North Sea oil was our rìposte to OPEC, a reinforcement of national pride and a reminder to the rest of the world that Britain was still in business. However, growing awareness of the difficulty of exploiting the oil reserves, and the realization that politically the North Sea was about to be nationalized under BNOC*, brought the public and eventually the press, back to reality. It became clear that, although there might be enough of it to replace imports and leave some over for export, North Sea oil would be dear, and had a life span which would bring it to exhaustion well within the lifetime of all those not already well advanced into middle age. North Sea oil has therefore passed quickly through its brief romantic period and taken its place alongside the various nationalized boards as a further reinforcement to the public sector. The more extreme promises held out by the first discoveries of North Sea oil and gas will not now be fulfilled. Nevertheless, Britain is more favourably placed with regard to energy supplies than any other European country.

In international terms the end of the oil era has been placed in the early 1990's, followed by perhaps a further decade of more or less constant production. Discoveries of new reserves or an improved recovery rate could prolong production, but only long enough to delay the need to switch over to other energy sources. For Britain the availability of North Sea oil postpones the moment of truth, for how long no one knows.

* The OPEC monthly bulletin advised members against investment in North Sea oil on the grounds of high costs (100 times greater than the 8 cents a barrel for production in Kuwait) and the fear of nationalization by BNOC. See *The Middle East*, August 1976.

Chapter 4
Obstacles to energy policy making

Not as simple as it looks

To the layman, the working out of a strategy for energy appears to be reasonably straightforward. The nation needs energy for heat, light, transport and to drive industrial machinery. If energy has to be imported, this means that foreign currency has to be earned to pay for it. Some of this will come from the export of manufactured goods, some from invisible earnings for insurance, consultancy services, banking, and so on. The higher the proportion of indigenous fuel used, the less the strain on the balance of payments. If we have enough fuel to be self-sufficient, so much the better. If not, then we must earn the money to pay for imports.

In practice the position is complicated by the peculiar characteristics of the fossil fuels – coal, oil and natural gas. Each requires considerable investment before production can begin, so that, although reserves may be known to exist, their development will only take place if supplies are likely to be sufficient in quantity and quality to justify the necessary investment. Each fuel has certain prime uses for which it is most suited and which it performs at maximum efficiency. To use fuels for other purposes is generally regarded as wasteful. Oil is exceptional in that it is the only fuel suitable for use in road vehicles and aircraft. All fuels make their characteristic contribution to the pollution of the atmosphere and the environment.

Energy policy involves a choice between the fuels available. This requires striking a balance between long- and short-term factors, which may at any particular time appear to be in conflict with each other. Market forces come into the account, but there is no sense in which prices are determined only by straightforward competition. There are different costs of production for the same fuel in different countries, and in different parts of the same country, as, for example, between the east Midlands and the south Wales and Scottish coalfields. In Britain the coal, gas and electricity industries are in public ownership while oil is still largely handled by private corporations. The nationalized concerns operate under rules which are a complicated balance of commercial and social considerations, related only indirectly to the market mechanism. Nationalized industries cannot, except in extreme cases, go bust.

Costs and prices in the energy industries are arrived at by regulation and control. Market prices are based on accounting costs, but these may appear to be out of line with the resource costs to the community as a whole. The need to avoid a rapid fall in employment in the coal industry in the 'sixties was an important factor in slowing down pit closures, and in keeping up the price of coal. For all fuels current prices reflect the long lead time on investment, the long pay-off period on research and development, and the inability of the fuel industries to react quickly to changes in the market. Another complication is the need to conserve reserves. The financial priorities – for example, the desirability of securing a high cash flow on a large investment project – may be possible only if resources are used up at a faster rate than is desirable from a conservation point of view. This may mean using one fuel for purposes for which another is better suited, as in the case of burning natural gas rather than coal in power stations.

Cheap energy policy

The dangers of a cheap energy policy were brought out by the build-up of events following the publication of the 1967 White Paper (Cmnd 3438) which advocated a four-fuel policy, leaving the market to regulate prices and production. This policy, which was in line with what was happening in other industrialized countries, led to a great increase in oil imports and placed OPEC in a dominant market position. While it is true that world demand for oil was rising, particularly in the US, as a result of increased growth rates, this did not justify the all-out concentration on oil to the neglect of other fuels. The considerable economies in the use of oil since the 1973 price increases show that the rise in consumption in the late 'sixties was not a matter of necessity but of convenience. The 1967 White Paper underlines another element in energy policy making, namely that at any given time the different fuels are competing with each other not only in sales and production, but for investment funds. This competition favours fuels which are in a position to supply the market at prices which cover financial and production costs, a position occupied by natural gas in the 'seventies.

Policy priorities

Energy policy priorities have changed completely since the 1973 oil price increases. It is not clear that a free market solution by itself will produce a relationship between fuel utilization and balance of payments considerations which optimizes the former and reduces the strain on the latter. The basic objective of energy policy is to ensure adequate and secure supplies of different fuels at the lowest possible real resource cost over the medium term. This broad definition needs to be spelled out in more detail. An adequate supply of fuel is one which imposes no constraints on either economic growth or the rise in living standards. Security, as we know only too well, cannot be proof against all hazards. It may be upset by political action, as with OAPEC oil, or by industrial action, as with coal in 1972 and 1974, or by accidents and natural catastrophes. An indigenous supply would generally be taken to be more secure than one wholly or partially dependent on imports, but this is not always the case. Again, the quality of an indigenous supply may not be so good as that of an imported fuel, or security considerations may result in using, say, indigenous coal instead of imported natural gas. Resource costing includes not only the cost of production of individual fuels but factors such as distribution costs, prevention of pollution hazards, environmental considerations and other values. The time element in energy policy introduces consideration of the relative merits of using a particular fuel now or later, and the arguments in favour of alternative courses of action. Securing an adequate fuel supply at reasonable resource cost involves the reconciliation of technical, financial and social problems, using a variety of estimating and analytical techniques, and checking these against each other.

What replaces competition?

The free market energy policy which operated largely *faute de mieux* from the 'fifties onwards, received its knock-out blow from the 1973 oil price rises. In Britain the theory was that coal, oil and natural gas competed freely against each other, while nuclear energy, because of its strategic importance, was developed by government decision without any close reference to market criteria. In so far as coal bore heavy social costs because of the need to keep uneconomic pits going in areas where coal mining was the only form of employment, it was not operating in a free market. Natural gas sales and distribution were under the control of the Gas Council (later Gas Corporation) which as sole purchaser was able to negotiate a low price for supplies from the producing companies. Oil was cheap because it reflected the low cost of production in the Middle East which enabled the oil corporations to market oil throughout the world at prices that were highly competitive with the price of coal. In other words, the 'free market' was in fact neither 'free' nor a 'market'. The NCB, burdened with the high labour costs of pits it would have preferred to close down, could not earn the money needed for large scale investment on rationalizing and streamlining the coal industry. Natural gas had the advantage of supplies at a monopoly price, while oil was produced by multi-national corporations on behalf of host governments that could not develop their resources themselves, and distributed through agreed arrangements. Two nationalized industries and seven of the world's largest multi-nationals hardly constitute a framework for free competition. In so far as this was a competitive system it was a highly imperfect one.

Electricity the policy regulator

The only point where anything like competitive conditions existed was in the sale of fuel to the electricity industry. So long as the demand for electricity was rising, as it did through the 'fifties and 'sixties, the industry expanded its generating capacity and replaced outworn plant. The electricity industry provides a secondary fuel, for the production of which it is dependent on coal, oil, gas, nuclear energy or hydro-power. In the 'fifties the rate of installation of new generating plant exceeded the rate of growth of the basic demand for electricity. However, the basic demand is the general average take-off which may be expected at ordinary times. There is also the peak demand which occurs when a number of factors, each making heavy demands on the supply, all operate together. The supply must be able to meet these peaks, which means that installed capacity is well above the basic demand.

The main cause of peaks in demand, reflected in changes between seasons, and between different hours of the same day, is the level of domestic consumption. The existence of a margin of unused capacity during off-peak hours has enabled the electricity industry to promote the sale of electricity for a diversity of domestic uses, including space heating by night storage heaters which store electricity generated at off-peak hours. In terms of fuel efficiency, the increased use of electricity is not an advantage when it takes the place of more effective primary fuels. However, from the consumer's point of view, electricity has the advantage of flexibility and the necessary appliances have low installation costs. Over the years the electricity industry has, except for short periods, been encouraged to burn coal and has received a subsidy to increase its 'coal burn' over and above what it would otherwise be.

As a secondary form of energy, the electricity industry has the choice of using different primary fuels in its power stations. In practice this freedom of choice may be limited by government decisions on the use of individual fuels. In this connection the concept of the optimum plant mix is important, so that cost differences can be evened out, and over-dependence on any one fuel avoided. An arrangement which enabled electricity to adapt to a multi-fuel policy would provide greater flexibility in fuel use. The only direct contribution that the electricity industry can make to the national energy resources is through the expansion of nuclear power production. Here the question of base and peak loading becomes important. Nuclear power could be developed for base load supply, leaving fossil fuels to meet peak requirements. However, investment in extra nuclear plant which involves displacing serviceable coal or oil fired plant is a waste of financial resources. The expansion of nuclear energy will therefore depend on the growth in the use of electricity, unless fossil fuels are going to provide decreasing shares of power station fuel in the future.

Peculiar position of electricity

According to CEGB forecasts consumption of electricity will stabilize at about its 1975 level, and then begin to rise slowly until the mid 'eighties. This increase is based on the assumption that there will be an end to the world wide recession and a resumption of economic growth. This increased economic activity will be associated with the development of North Sea oil. In the short term the CEGB will be meeting the demand for electricity using its existing generating plant following a merit order which will give the highest thermal efficiency. In 1975-76, the CEGB used 67·5 million tonnes of coal, an increase of 2·7 million tonnes over the previous year, in spite of reduced electricity demand. However, fuel consumption at 9·7 million tonnes was the lowest since 1969-70, in spite of the completion of new 2000 MW oil fired power stations at Fawley and Pembroke. Interruptible supplies of natural gas amounted to 3·8 million tonnes coal equivalent. These supplies consist of gas which has to be piped ashore in the course of regulating the supply system. The alternative to disposing of a limited quantity of gas in this way would be to build very large, and costly, storage capacity.

The electricity supply industry has a unique position in so far as virtually every household and business premises in the country has a mains supply of electricity. This is not true of the primary fuels coal, gas and oil. Also, under present statutes the electricity industry is obliged to supply the consumer, which is not the case with the other fuels. This means that the demand for electricity is uniquely dependent on the supply and demand relationship for each of the other fuels. At the same time, the supply of electricity is directly dependent on the supply of each other fuel. Electricity is expected to be available at any time and, because it cannot be stored, sufficient capacity has to be provided to supply peak period demand.

Electricity is in competition with its own suppliers. Coal, oil and gas are all used in electric power stations, and 51 per cent of operating cost is absorbed in their purchase. If oil or gas or coal oust electricity from one of its main markets – e.g., for cooking or domestic space heating, or industrial purposes – the demand for electricity would fall. There would also be implications for the demand for the other fuels used in the generation of electricity. Competition from gas or oil in space heating could reduce the demand for coal for electricity. Paradoxically, there could be a situation in which sales of coal for, say, metal melting could damage the market for coal for electric power stations. The energy market, never simple, is uniquely affected by both losses on the roundabouts and profits on the swings.

Future prospects

Electricity supply emerges as an important key to the form of future energy policy. The possible exhaustion of oil and natural gas in the next quarter century means that they should be put to prime uses and not burnt in the boilers of power stations. Nuclear energy can only be used for the generation of electricity, which is also the principal use to which coal is now put. Looking ahead, coal will be used increasingly as an industrial feedstock and for the production of substitute natural gas. The competition between electricity and gas, oil and coal for space heating is not a real issue. So far as can be seen, space heating, industrial power and, of course, lighting, will be provided increasingly by electricity as gas and oil are depleted. Conservation policies should aim to concentrate the use of available oil supplies as fuel for transport. Natural gas, to be replaced eventually by synthetic products made from coal, would be the great stop-gap fuel of the next 20 years or so. This does not mean that it should be used wastefully, but regarded as an additional plus factor in securing faster growth and moving into surplus on the balance of payments. In other words, long-term policies for coal and nuclear energy would be related to the future level of electricity supply and use, and in the case of coal to its use as a feedstock. Oil would be concentrated on producing petrol, jet fuel and diesel with residual fractions for chemical feedstock. Natural gas is the only indigenous fuel that is in any true sense expendable, as substitute gas will take over its major uses.

Four-fuel summary

Although there appears to be a choice between fuels, determined by price, convenience and above all past investment in appliances and equipment, this is only a short-term position. In the long term, certainly within the next century, oil and natural gas will run out, leaving coal for maybe a further hundred years and, if uranium supplies can be organized, nuclear energy. Each of the four fuels has its own prime uses, but the four have one major use in common, the production of electricity. Nuclear power can only be used commercially to generate electricity. Coal and natural gas have broadly the same range of uses but gas is a more convenient, cleaner and cheaper fuel. Also, the cost of production of natural gas, once it has been discovered and brought on stream, is well below that of labour-intensive coal. Oil can be used for steam raising and in power stations but is unique as the fuel for land, sea and air transport.

The connecting link between the four fuels is electricity, which transforms fuel burned in power stations into energy for power and light, which can be made available almost anywhere. Coal is the principal fuel used for this purpose, and the electricity boards are much the biggest users of coal. A decreasing amount of oil is used in power stations, and there is resistance to using gas, as it can be used much more efficiently for direct space heating, heat processes and so on. All the nuclear power so far available goes to the power stations.

Competition between fuels

The points at which competition takes place between fuels are, first, in producing electricity in power stations and, second, between coal, gas and oil for domestic space heating and industrial high temperature processes. Because of the need to maintain additional capacity to meet peak demand well above basic load, the electricity authorities can offer cheap supplies for some purposes. Competition is on the basis of complicated price structures related to existing production costs and to the historical cost of the investment on which the supplies depend. In the short term competition is concerned with pricing the different fuels according to demand for potential supplies by domestic and industrial consumers. In the longer term account must be taken of the investment needed to provide supplies in the quantities calculated to be necessary. Consideration of both elements is necessary to the formulation of energy policy. The position is complicated by the fact that both supply and demand are composite and the constituent parts are not fully interchangeable. Also, energy policy is only a part of economic policy. It must fit into policies for total domestic investment, and not upset the balance of payments. Many of the oil importing developing countries have found their forward planning completely upset by the disproportionate amount of their resources that has to be devoted to providing for even minimum fuel requirements in conditions of inflation.

Difficulties over priorities

If competition is to regulate the production and use of fuels, a large number of conflicting priorities have to be assessed. It is generally agreed that some way of equating the decisions affecting different fuels must be found as a basis for competition. The economic prescription for this requirement is that different fuels should be sold at prices equated to long-run marginal costs.

Comparisons of fuel prices are always difficult to make because published prices are subject to unspecified and sometimes substantial discounts. A further difficulty is the fact that prices are quoted in different units and are not readily available in a form that reduces them to a comparable base. The fuel policy White Papers of 1965 and 1967 did not mention the price of a single fuel. The 1967 paper made various generating cost comparisons between nuclear and conventional power stations, which were favourable to the AGR stations as compared with new coal and oil fired plant. For every fuel there is a range of costs, each associated with a particular source of supply. The Electricity Council has coal, oil, gas and nuclear fired power stations, each type of which contains a variety of different examples of plant, using fuels at different costs. The nuclear power stations include Magnox stations, now over 20 years old, and the Hinkley Point AGR station, which came into use during 1976. The coal fired stations receive coal of varying qualities and production costs. Differences also arise in the cost of fuel oil from different sources, with different transport and refining costs. In short, energy policy consists in reconciling a series of average costs, administered prices and ranges of potential demand and setting these against potential supplies. When various other factors, such as the tax on imported oil, taxes on petroleum, the level of taxation on industry and the individual consumer, are fed into the pricing programme and account is taken of the past and prospective investment programmes of energy suppliers and consumers, the dangers of over-simplification became apparent.

Ground rules

With these various factors in mind, it has been proposed* that instead of indulging in long-term forecasts governments should draw up a set of

* Sir Peter Menzies, 'Flexibility in long-term energy policy', Electricity Council, 22 June 1976.

ground rules which all energy industries would adopt as the basis for their future development. These would be concerned with safety and the preservation of the environment; the rate of return required for the production, distribution and utilization of energy; principles for pricing and methods of payments for energy; the degree of freedom of consumer choice for fuel; ownership and control of energy sources. The difficulties with such a set of rules would be, first, in securing agreement on their provisions and, second, in their interpretation. The drafters would have to devise an energy policy, based on forecasts and scenarios, to meet the agreement of the interested parties, and then reduce this to a set of ground rules. Flexibility of this kind would be the ideal arrangement, provided all the interests concerned could agree. In present circumstances it is possible to imagine a set of ground rules forming a part of energy policy, but only if it is backed up by a formidable set of legal powers and sanctions. Every decision involving the public sector has its non-economic aspects.

The new interventionism

Although competition between fuels is still talked about in terms that would have won Adam Smith's approval, the fact remains that the past 30 years have seen increasing intervention in the operation of the private sector. In the decade following the end of the Second World War, governments were mainly concerned with establishing full employment and economic stability on Keynesian lines. However, the introduction of counter-cyclical policies depending on the operation of fiscal and monetary measures soon proved to have limitations. In particular, no readily applicable solutions were forthcoming for the problems of declining industries and depressed regions. At the international level, the management of world currencies and trade proved equally difficult, particularly in the case of the overburdened, post-imperial, British economy. Of especial concern was the rapid build-up of inflation to near-crisis proportions in the mid 'seventies.

Within this unpromising international framework, governments have attempted with mixed results to introduce some type of formalized planning. The French Commisariat au Plan was taken as the model by the Macmillan Government when it set up the National Economic Development Council (NEDC), generally known as 'Neddy', in 1962. The Labour Government of 1964 went farther and created the Department of Economic Affairs, which was in effect a planning ministry. The difficulties of stimulating economic growth to reach official planned targets, while deflating the economy to maintain the sterling exchange rate, not surprisingly led to the abandonment of the first National Plan in 1966. However, successive governments have held to the view that every economic problem requires a political solution. Leaving aside the detailed measures introduced to help particular industries or regions, to control wages and prices, and to regulate the activities of trades unions, interventionism has had a number of successive phases. Some of these began with the enunciation of an idea – for example, Mr (now Sir) Harold Wilson's 'white heat' of technology campaign which was intended to raise productivity and bring faster growth to close the technology gap. Others have centred on a particular action – for example, the decision to fix a £6 a week maximum for wage increases for 1975-76 – and treat this as though it was a complete economic policy in itself. Other initiatives have been limited to particular projects, generally involving state control and finance. These have included the development of the 'Concorde' aircraft and the AGR second-generation nuclear energy programme. In these cases, although private sector firms were engaged in some parts of the activity, the planning and financial arrangements have been in the hands of civil servants. Governments have, in fact, moved from the position of providing broad guidelines for economic growth, to the running of a whole string of nationalized industries and the master-minding of specific projects. Ahead lie new areas of intervention in external relations with growing government regulation of foreign trade. The concern about the market shares of industries like textiles and Leyland and Chrysler cars is, perhaps, an indication of the pattern of future state activity. Under the present system, with ministers responsible for implementing policies requiring executive decisions, there is the danger that economic policy will end up trying to give everything top priority. State intervention, once begun, spreads steadily outwards as the case of the Drax B power station demonstrated. The problem here was to preserve the jobs of the workers in the two turbine generator manufacturing firms, GEC and C.A.Parsons. The fact that the CEGB did not require additional capacity at the time was regarded as a side issue.

Intervention and energy

An important new factor in energy policy is the growth of state intervention in the oil industry. In the past, governments intervened to supervise prices and investment in the nationalized industries with the object of ensuring that financial 'targets' would be met and supplies forthcoming in line with demand forecasts. Nuclear energy was exceptional in the sense that a commercial return on investment was not expected in this case. The fact that it did not always materialize in the case of coal, gas and electricity was not taken as a sign that these industries were 'failing the nation' but of the special difficulties of the mixed economy. The oil industry, by contrast, was left largely alone except for the imposition of the tax of 2·0 old pence a gallon on oil, amounting to £2 a ton. Until the discovery of North Sea oil there were no indigenous resources to develop, and British governments were content to leave the industry to the major oil corporations. The only departure from this policy was the decision to encourage domestic refining instead of relying on imports of petrol and other refined products. The freedom of the oil corporations to invest and explore in the North Sea has been qualified, but the desirability of accepting foreign private investment and technology is recognized. In the new situation the cost of imported oil quadrupled since October 1973 and it is generally realized that world oil and gas reserves have only a relatively short life. In considering intervention in the energy industries, therefore, we are not concerned only with the free play of traditional market forces, but with the conservation of scarce resources over a period of uncertain duration.

Yet another ingredient in this formidable mixture of problems are the hazards of party politics. These limit the freedom of action of partisans through commitment to past pledges and interests, but make certain courses of action impossible without the consent of important sections of the major parties. What is proclaimed as a national energy policy may turn out, on examination, to be a compromise arrangement designed to satisfy as large a proportion as possible of the supporters of the party in government. This may be the only possible political answer to the question, but it is wishful thinking to suppose that it is likely to lead to a fundamental solution. In the modern world, variable factors are more numerous than before. This is especially true of the energy industries, where the experts have been consistently guilty of misjudging the future pattern of supply and demand, governments have in consequence been led into ill-starred decisions. The economic system is not a piece of machinery, but an environment, a set of relationships between human agents and natural forces. While increasingly aware of the difficulties of predicting the future, we cannot remain indifferent to it.

Chapter 5
Forecasts, scenarios and policies

The seaweed syndrome

Thanks to the modern paraphernalia for gathering and processing data, we know a great deal more about the future than did previous generations. Unfortunately, it is not possible to say with any certainty that what we assume will turn out to be right. All the best weather forecasters are said to keep a piece of seaweed handy and to check their scientific conclusions by its behaviour. There is nothing similar for economists to fall back on. They have to rely on extrapolations of past trends, scenarios designed to fit different sets of possibilities, and the ability to sense changes in the wind at an early stage. If all that was required of them were generalized statements that 1978 would be a bad year for commodities or that a major financial institution would be in trouble early in 1979, on the lines pioneered by Old Moore, economists could sleep soundly at nights. As their forecasts are used as a basis for policy decisions determining national expenditure on consumption, investment and welfare, an element of realism is desirable.

In the field of energy, with its long lead times, uncertainties on the size of fuel resources, possibilities for technological error, and all the hazards of fuel politics, forecasts rarely turn out as well as their authors expect. There are those who argue that economic forecasting is, in any case, a vain conceit and that supplies and prices should be left to be determined by the working of the market. On this assumption fuel prices would respond to changes in demand, so that if more electricity was consumed the prices of coal and oil for power stations would rise, and presumably vice versa. In practice, with the various fuels, except for the moment oil, under state control, prices are most likely to go up in conditions of falling demand in order to cover costs from lower sales. Nationalized monopolies in the energy field and elsewhere are not particularly sensitive to market forces.

Priorities and premises

Whether we believe in long-term forecasts or not, a number of key issues have to be considered. The first is quite simply, do we need to be self-sufficient in energy? At present rates of depletion, lead, zinc, tin, platinum and gold, to mention only a few of the more important minerals, will be exhausted before coal and probably oil. Should we therefore begin worrying about policies to make ourselves self-sufficient in tin, or zinc by 2000 AD? So far this eventuality has not been covered in the manifestos of the main political parties. Again, if Britain becomes self-sufficient in energy in the 'eighties, thanks to North Sea oil, it will be so for the first time since the great days of steam power over a century ago. And it will be a condition attained at the cost of enormous investment crowded into a brief period of time, and undertaken largely for the wrong reasons. At present there is a positive glut of oil, so that the immediate reason for developing North Sea resources is not to secure supplies but to reduce the balance of payments deficit. Conservationists complain that North Sea oil is being developed too soon because OPEC oil is available on world markets. This time and money could be better used to develop solar, wind, wave or some other alternative sources of energy against the day when OPEC oil supplies begin to dry up. This would do nothing to help the balance of payments in the immediate future, however, so under present policies Britain's oil reserves will be used up, and research on alternative sources of energy will have to wait. On the question of security of supplies the Electricity Council has somewhat unkindly remarked that indigenous coal has in recent years been much less secure than imported oil.

Energy prices

For energy, perhaps even more than for other commodities, price is only one of the factors taken into account by the consumer. More important is the capital cost of the equipment required, the running cost, the reliability and convenience in use of individual fuels. Gas is highly popular for space heating not only because, temporarily, it is cheaper than oil, coal or electricity, but because it is convenient to use. For the present, consumers are free to choose which fuel they will use, provided they are willing to undertake the necessary expenditure on installing the required equipment. On this basis the prices of individual fuels, provided they do not vary too widely, are not the decisive factor in determining choice. On the whole, prices should reflect the costs involved in producing a fuel and bringing it to the point of use. But if the costs are of different orders of magnitude, the price for a particular fuel may be so high that consumers could not afford it even though it would be their preferred fuel. Significant price differences can raise the political problem of 'rationing by price', which some regard as undesirable. The alternative to charging prices which reflect costs, even if this means high prices for some fuels, would be rationing by administrative action. This could take the form of an allocation of a certain quantity of fuel at a special price to particular categories of consumer, such as pensioners, or it could mean allocating fuels to particular uses. Alternatively, fuel prices could be fixed arbitrarily by the authorities so that some fuels might be sold at a loss and others well above costs. Most forms of price fixing require large numbers of officials to administer and police them. If it is necessary to ease the burden of fuel costs for the needy, then ways of doing this can be found by increasing pensions and social benefits rather than by subsidizing energy prices.

If energy prices are to be related to production costs and not fixed arbitrarily, they should be equated as far as possible to long-run marginal costs. This involves a calculation of the cost, starting from now, of an additional increment of supply on a continuing basis, taking into account the discounted capital cost and operating cost of all the plant required to do this. More precisely, for electricity supply 'marginal cost for any year is the excess of: (a) the present worth in that year of system costs with a unit of permanent output increment starting then, over (b) the present worth in that year of system costs with the unit permanent output increment postponed to the following year'.*

* A. Turvey, *Economic Journal*, June 1969.

Measuring energy consumption

Energy consumption can be measured in three different ways. The first is the input of primary fuels needed to produce the energy. The second is the heat supplied by the primary and secondary fuel industries after allowing for conversion losses. The third is the useful energy remaining after allowing for losses in appliances. All three are measured in terms of millions of tons of coal or oil equivalent, or in billion therms. The distinction between the three measures of energy consumption is important because, depending on the efficiency with which they are used in conversion processes or final use, the same amount of primary fuel or heat supplied can yield very different levels of useful energy. Changes in efficiency need to be taken into account in energy forecasts. Fuel saving through greater efficiency in use is an integral part of energy policy along with the development of new sources of supply.

The relationship between energy demand and price also creates problems for forecasters. The general level of energy prices affects the total of demand, while the relative prices of individual fuels may be of considerable importance in deciding the mix of fuels which goes to meet total demand. At any particular time the energy market will be affected by the physical availability of fuels and the inevitable time lag in meeting changes in demand. When oil prices rose above the average costs of competing primary fuels, substitution of other fuels for oil was slow to take effect and has been limited to increased gas consumption in the domestic market and in industry, with more coal being used in power stations. The continuing high price of oil has stimulated the development of additional supplies of other primary fuels but the effect of this is not noticeable in the short term. In considering fuel prices to the end of the century it is better to concentrate on relative prices rather than to attempt to predict absolute levels for each fuel.

The Department of Energy view

The views of the Department of Energy on Britain's energy supply and demand were set out in a statement for the National Energy Conference of June 1976. Besides making assumptions about the growth of GDP as a whole, the Department considered what might happen to the growth of consumer expenditure and industrial production, both of which affect demand for energy and the requirement for individual fuels. The relationship between economic activity and energy consumption is not constant and both higher prices and advances in technology are generally expected to lead to the more efficient use of fuel.

In a review of individual fuels the statement concluded that the output of coal after 1985 would be limited only by the rate of new production achieved to offset the exhaustion of existing pits. Costs are likely to rise in the older, and heavy capital expenditure will be involved in the development of new, pits. However, new and reconstructed capacity will have a higher level of productivity and for the industry as a whole productivity will rise as new capacity is commissioned. The estimate for natural gas reserves on Britain's Continental Shelf at the end of 1975 was some 29 000 trillion cubic feet of proven reserves with a further 22 000 trillion cubic feet as probable or possible in known discoveries. These figures are regarded as cautious and do not include possible imports from the Norwegian sector of further amounts of gas to be discovered possibly in association with oil. The forecast is that gas will be used up rapidly to about 1980 in order to displace costly imported oil, after which a more conservationist approach would be adopted and supplies would go to premium demand in the domestic sector and for special processes in industry only. Slower growth in total energy demand and the availability of more associated gas could mean that larger supplies than anticipated would be available in the last decade of the century.

At present nuclear power stations provide about 4·3 GW of generating capacity representing 4 per cent of total energy supplies, but 11 per cent of total electricity supplies. When the AGR power stations, whose construction has been delayed for various reasons, come into service, the total nuclear generating capacity will be about 9·4 GW producing 8 to 9 per cent of total energy and nearly 20 per cent of electricity. The programme of thermal reactors, originally based on SGHWR reactors scheduled to be commissioned in the mid 'eighties, would have brought nuclear capacity up to about 11 per cent of the total energy and about 22 per cent of electricity production. Research and development programmes now in being could result, if actively pursued, in the fast breeder reactor being available on a commercial basis after 1990.

Oil will continue to be the essential fuel for transport, as a feedstock, and for many industrial uses. Net imports are expected to continue to 1980 and for marginal requirements only after that date. North Sea oil, it is assumed, will be priced at its full opportunity cost on the basis of world prices. It is estimated that supplies will exceed present forecasts of oil consumption by 1980 and that this will continue possibly for the next decade, although by the mid 'eighties production will depend increasingly on new developments and discoveries, and on the depletion strategy followed.

The Department of Energy statement concluded that the combined effect of price increases, economic recession, and increased gas supplies could severely reduce the electricity demand on the domestic market up to 1980. Total electricity demand is expected to rise, but at a less rapid rate, to the mid 'eighties. If the supply of gas begins to taper off, electricity sales could expand rapidly. The average rate of plant ordering for power stations is likely to be low to 1980 as demand remains depressed, but is taken to increase sharply after that, to an installed generating capacity by 1990 of somewhere between 15 and 30 GW. Present lack of demand from its principal domestic customer is causing considerable difficulty for heavy electrical plant manufacturers. The choice of fuel will depend on relative fuel prices and the capital cost of the plant, and considerations of flexibility and security of supply. The construction of coal and oil fired plant is likely to be affected more by the general economic situation than by relative fuel prices. It is expected that coal will be the fuel used in new power stations. The construction programme for nuclear capacity is not likely to be affected by changes in the economic situation or in the price of oil. The supply and demand situation for the four primary fuels is shown in Table 5.1. From this it will be seen that a deficit of 149 million tons of coal equivalent in 1975 is transformed into a surplus by 1980 of 15 to 30 mtce which tails off thereafter to an amount depending on how far the level of production of North Sea oil rises above 165 mtce.

Economic growth and energy demand

With a 2 per cent annual growth rate primary fuel consumption which now stands at about 340 mtce could reach 400 to 450 mtce in 1990 and would rise to about 550 mtce by the year 2000. As electricity consumption will be growing more rapidly within this total its share would rise from about a third of total consumption in terms of primary fuel inputs in 1975, to over a half in 2000. However, bearing in mind earlier forecasting excesses, a more cautious view would be that a 2 per cent annual growth rate in primary fuel consumption probably represents

Table 5.1

UK primary fuels: supply and demand 1975-90
(million tons of coal equivalent)

	1975	1980	1985	1990
Energy uses				
Coal	120	120–125	125–135	115–150
Gas*	50	65–75	65–95	65–95
Nuclear and hydro	13	25	30	40–55
Oil	133	100–125	100–165	115–225
TOTAL ENERGY	316	325–345	350–400	395–465
Non-energy oil and gas	24	35–40	40–45	50–55
TOTAL PRIMARY FUELS	340	360–385	395–445	440–520
INDIGENOUS PRODUCTION	191	375–415	390–505	390–555
of which North Sea oil	1	160–190	165–250	165–250
Surplus or deficit	−149	+15 to +30	−5 to +60	−50 to +34

* Excluding non-energy uses.

Source: Department of Energy.

an upper limit, and that the combined effects of higher prices, market saturation and fuel saving policies would result in a slower growth in demand for electricity. As increase in demand may become more difficult to meet, prices are likely to increase and their effect could only be offset by conservation policies with higher standards of efficiency from lower energy inputs. Because of the long lead times the extent of improvement in performance depends on investment to be made in the next few years.

On the basis of present knowledge, oil from the North Sea would be approaching exhaustion before the year 2000 but demand for oil would continue to rise in the transport sector unless major technological changes, for example in the development of electric vehicles, have been achieved. On the same assumption, natural gas would also be in decline, limited in its use to premium markets and supplemented by costly substitute natural gas made from oil or coal. Further discoveries of oil and natural gas could delay the arrival of this situation, but a more likely break-out would be through the development of coal production and new conversion processes. Electricity, assuming a growth rate twice that of energy consumption as a whole, could represent well over half total primary fuel demand by 2000. Coal and oil would be in greater demand for non-energy uses by that time and their contribution to electricity generation would have fallen. Nuclear energy could be making an increasing contribution to electricity supplies, possibly with a growing element of fast reactors to enable efficient use of uranium reserves. This is assuming that the difficulties over the development of the fast reactor will have been overcome. A small amount of electricity may be produced by unconventional sources such as wind, waves, and solar energy.

This official view of the way in which supply and demand for fuels is likely to develop until the end of the century assumes that Britain will enjoy a period of relative abundance to the 'nineties and perhaps beyond. This will provide surpluses for export, and decisions will have to be taken on how much to export and how much to conserve. However, although plentiful, fuel will not be cheap. The existence of ample coal reserves will provide the basis for a continuing energy policy, but the indications are that Britain will again be importing costly energy before the end of the century. Provided the economy is sufficiently strong to generate the finance to pay for imported fuel and coal, and nuclear energy has been developed as planned, fuel imports should not present impossible problems.

The general conclusion is that by 2000 we shall be back where we started from, with an energy gap and the windfall resources of natural gas and oil nearing exhaustion. By that time new sources of energy may have been developed, the fast breeder reactor may be in operation so that the dangers of a shortage of uranium would have been offset, and new technologies may have reduced the importance of oil for air, land and sea transport. However, it requires considerable optimism to assume that all these favourable events will have happened. When we look at the nuclear energy industry and consider that it is based on research that was going on in the 'thirties with practical application to nuclear power stations in the early 'fifties, and consider the false starts and problems that have arisen since, it becomes clear that a long time-scale is required for new sources of energy to be developed. Bearing in mind also that the Department of Energy survey is based on alternative growth rates of GDP of 2·5 and 3 per cent (taking 1974 as a base) it is possible that these growth rates may not have made sufficient allowance for continuing inflation. For the world as a whole, an increase in the use of energy by the developing countries could well outstrip any savings by conservation in the industrialized states.

Some scenarios

Past failures of attempts to forecast energy demand by extrapolating historical trends into the future have led to the introduction of less conventional forecasting methods. A variety of scenarios designed to produce a strategy for different demand and supply patterns have been produced. Of these the most interesting in many ways are the seven scenarios prepared by the Advisory Council on Research and Development for Fuel and Power (ACORD).* The purpose of the ACORD report was to provide a framework for the future planning of research and development in energy technologies. The report accepts the general view that about the turn of the century a widening gap between Britain's overall demand for energy and her indigenous supplies of fossil fuels will develop. ACORD is concerned with the need, in view of the long period required for the results of research and development to take effect, to examine the contribution that R & D programmes can make to bridging this energy gap. The scenarios identify the most important technologies required in possible future conditions and the R & D strategies that each would require to be embarked upon now.

The do-nothing-new scenario

The scenarios have been chosen with a view to ensuring that future policy makers have available appropriate energy technologies to implement their policies, whatever these may be. The report therefore takes an analytical approach and assesses technology needs over a range of plausible futures. The starting point is a trends-continued view which extrapolates middle of the road views of likely social, economic and technological developments in the medium term through to the

* 'Energy R & D in the United Kingdom', a discussion document, ACORD, HMSO, June 1976.

year 2025. Given an economic growth rate of about 3 per cent and assuming energy consumption to be limited largely by price, Britain would be able to maintain self-sufficiency until the end of the century. The nuclear building programme is assumed to be geared to the installation of 50 to 60 GW(e) by that time. After that, with the expected decline in offshore oil and gas, and with the coal industry producing 150 million tonnes a year, an energy gap which could be filled only by nuclear energy or imported oil and natural gas would open up. Britain would be importing fuel at a higher rate than ever before, unless nuclear capacity over 100 to 120 GW(e) was installed by 2010. Shortages of uranium would almost certainly necessitate the introduction of fast reactors by the late 'nineties. All things considered, the trends-continued, do-nothing-new scenario is not particularly encouraging.

Variations on a theme

The first scenario takes a low-growth view and assumes continuing international economic recession with British economic growth of less than 2 per cent a year until the end of the century and under 1 per cent after that. Given a moderate conservation programme and an expanded coal industry it should be possible with low economic growth to manage with 75 (GW)e of nuclear plant when North Sea oil and gas give out. In this case fast reactors would not be needed so soon to meet the shortage of uranium, but could be required for quite different reasons.

Scenario 2 takes a situation in which nuclear power programmes at home and abroad are abandoned in the early 'eighties by international agreement. At that point the SGHWR programme (or its successor) should have been completed. In this case, even with expanded coal production and strict conservation measures, the ending of oil and gas supplies would mean it would not be possible to sustain annual economic growth rates of even 2 per cent and economic standards would have fallen considerably by the end of the century.

The third scenario takes a high-energy cost view, which assumes a further sharp rise in world energy prices before 1980. Conservation measures and high prices would limit demand and ensure self-sufficiency well into the next century with British economic growth rates of up to 2½ per cent a year. In this case nuclear power stations giving 40 to 50 GW(e) in the year 2000, and twice that by 2010 would be needed. Fast reactors would be necessary on a large scale during the first decade of the next century.

The fourth scenario assumes energy prices to be in transition. This means that they would continue at their present levels until pushed up by a sharp, as yet unexpected, rise in world energy prices, possibly in the early 'nineties. As total world and UK demand for energy would have increased by a half and a third, respectively, by that time, a serious crisis would develop. Over 100 GW(e) of nuclear plant would need to be installed by 2000 for imports of fossil fuels to be kept below a level of 100 million tons of coal equivalent.

The fifth scenario takes a self-sufficiency view based on the acceleration of the nuclear building programme. This supposes an annual economic growth rate of 3 per cent requiring over 100 GW(e) of nuclear plant by the turn of the century. Shortage of uranium would make this programme impracticable if fast reactors were not operating on a large scale by the year 2000. In this case conservation measures would be taken to reduce consumption of fossil fuels, and the coal industry would be expanded to provide feedstocks for chemicals and substitute natural gas.

The sixth scenario takes a high-growth view which assumes a UK growth rate averaging around 4 per cent a year up to the early part of next century. Considerable difficulties would arise in securing fuel to meet rapid economic development on this scale. Without vigorous conservation measures to contain energy demand, self-sufficiency would not be possible after the early 'nineties. Fuel imports would exceed present-day levels even if 100 to 110 GW(e) of nuclear plant were installed by 2000 and double that amount by 2010. Such a programme would require the ordering of another series of thermal reactors in the early 'eighties and fast reactors would have to be in large scale operation by the mid 'nineties. In other words, a high growth rate would mean rapid depletion of oil and gas, and a much larger investment in nuclear energy at an early stage.

Importance of coal and nuclear energy

From this brief summary it is clear that whichever of the ACORD scenarios actually came to pass Britain would need a well developed coal industry and extended nuclear programme. Even a continuation of present trends would require almost 25 times the existing installed nuclear capacity if substantial energy imports were to be avoided around the year 2000. This would be the case even though the coal industry would by then have reached the *Plan for Coal* target of 150 million tonnes a year, which is 25 million tonnes above present levels. The reason for this is the growing need for coal to be used as a source of gas, liquid fuels and chemicals. The development of North Sea oil and gas is extremely important in the medium term; beyond that the more difficult technologies needed to develop deep water (over 200 metres) oil and gas will become increasingly important. The lesson emerging from the ACORD analysis is that complacency about future needs based on adequate medium-term supplies from the North Sea could delay investment decisions in nuclear capacity and coal until it is too late for them to be implemented in time to meet future requirements. The conclusion of this report, which is in broad agreement with the WAES report on the global situation, is that the key investment decisions concern the extension of nuclear capacity and the development of the fast reactor.

Conservation measures

Experience since the 1973 oil price increases has shown that considerable saving of energy is possible without reducing the pace of economic development. However the term energy conservation carries a variety of meanings, of which the most usual are thrift, economy and efficiency. The conversion of primary energy for use in a more convenient form – for example, burning coal to produce electricity – results in considerable losses. Improvements in the overall efficiency of these processes would be of continuing importance. Other losses occur in refining crude oil into petroleum products, an area where considerable savings could be made. Improvements in the electricity grid and in the capacity of gas pipe lines would also reduce energy losses. In other cases, the effectiveness of conservation measures depends on energy prices and the availability of alternative systems. District heating is an example of this kind of situation. Conservation measures in buildings, industry, and transport offer major opportunities for saving fuel, but are difficult to put into operation, as they depend on a multitude of individual decisions. The introduction of new building regulations, speed restrictions on transport, and improvements in equipment and processes would all help in this direction. Increased awareness of the importance of conservation could lead to more effective government support of R & D programmes, a variety of which are sponsored by the Departments of Industry, Energy and the Environment. In addition, substantial international support for R & D comes from the EEC, the International Energy Agency (IEA), the OECD and the NATO Committee on Challenges to Modern Society (CCMS). The CBI and the Department of Industry and Energy jointly sponsor the Committee on Studies Leading to Industrial Conservation of Energy (SLICE) which is undertaking a number of studies on the use of energy in industry. The programme is being managed by the Energy Technology Support Unit (ETSU) at Harwell, using consultants, research associations and individual firms under contract. Various specialized bodies are also involved in conservation research, notably the Building Research Establishment (BRE).

Coal research

Individual fuel industries have been carrying on their own research programmes for a number of years. The increase in the price of imported oil led to a major reappraisal of the role of coal in the British economy, as a result of which research and development effort has been increased and directed to raising the efficiency of coal mining, improving existing coal processes, and exploring new markets for coal and coal derivatives. This research is undertaken principally by the National Coal Board at the Mining Research and Development Establishment (MRDE) at Bretby in Staffordshire, and at the Coal Research Establishment at Stoke Orchard in Gloucestershire. In 1975-76 NCB expenditure at these and other establishments amounted to over £15 million of which about £2 million came from the European Coal and Steel Community. The budget of the MRDE was about £7 million in 1975-76 and the Establishment employed some 400 qualified scientists and engineers. Improvement of the efficiency of the longwall method of coal extraction is the main activity at MRDE. Over 90 per cent of deep-mined coal production in Britain is mechanized, with power loaders and armoured flexible conveyors, and powered support installations at the coal face. The NCB also have as a subsidiary and self-financing contract R & D organization, BCURA Ltd at Leatherhead, which is engaged in char gassification for the American Cogas

Development Co., and fluidized bed combustion for both British and American interests.

The ACEC projections

A report commissioned by the Department of Energy from the Advisory Council on Energy Conservation (ACEC), and prepared by the Energy Research Group at the Cavendish Laboratory. Cambridge,* takes a different line from the ACORD report. Two projections were made for energy to 1985 and 2000. The first of these took a high economic growth rate averaging 2·9 per cent annually from 1972 – 2000, and the second assumed a rate of only 1·9 per cent. In each case the energy coefficient (the ratio of the energy growth rate to the economic growth rate) was calculated.

The energy coefficient makes it possible to measure the rate of change of national energy efficiency in economic terms. If the energy coefficient is less than unity for a period, this means that 'economic' efficiency in the use of energy is increasing, although it does not imply that the level of economic efficiency in energy use is necessarily high. The Cambridge report showed the energy coefficient to have fallen to 0·3 in the late 'fifties after which it rose through the 'sixties due to special influences in the housing sector giving increased efficiency in energy use. The main reasons for this were the replacement of older, solid-fuel-burning, open fires by various forms of central heating, and improved insulation and draught-proofing. As households consume nearly a third of total energy, these changes were important in reducing the energy coefficient in the decade to 1968. In the following five years, energy required for the household sector increased appreciably, largely because of increased use of electric space heating. Higher energy prices and increased awareness of energy conservation measures are likely to be associated with lower energy coefficients in the years ahead.

The ACEC report described two projections for energy demand to 1985 and 2000. In 1985 there could be a substantial net export of energy from Britain following the rapid development of North Sea gas and oil supplies. Britain would be in energy balance apart from oil exports. The report warned that there may be difficulty in utilizing

* 'Energy prospects', Advisory Council for Energy Conservation, Paper Three, Department of Energy, HMSO, London, September 1976.

annual coal production planned at 140 million tonnes, unless some new coal fired power stations are constructed or converted from oil to gas. For the year 2000, projections which took both high and low growth rates and linked these with assumptions of high and low supplies of indigenous energy indicated a net import requirement in all four cases. The projections indicated a rapid transition from an energy export potential in 1990 to an energy import situation in 2000. World energy supplies were assumed to be increasing too slowly to match earlier trends in demand.

It is not clear whether Britain could maintain steady economic growth under these conditions, unless policies to reduce some of their possible adverse effects could be put into operation in advance. The ACEC report concluded that for the short to medium term, future British energy prospects are favourable compared with those of many industrialized countries. However, ACEC agreed with the ACORD report that there would be energy problems in the long term unless there is adequate development of the coal and nuclear power industries, together with vigorous measures for energy conservation. Because of the long lead times, sustained programmes must be put into operation during the medium-term future while Britain may be a net exporter of energy. The report stated that the government could influence several areas relating to future energy prospects. These included development of the coal supply industry and of more efficient techniques for the direct use of coal to enable it to compete more effectively with oil and gas; control over the rate of development and depletion of North Sea oil and gas, the preparation and implementation of the nuclear energy programme, and the launching of a comprehensive policy for energy conservation. In the long run the cheapest and most easily accessible energy source may be the energy saved by conservation measures.

The future waits

As the best-endowed EEC member so far as energy is concerned, Britain has done remarkably little about sorting out its policy objectives. The choice is difficult. Should we go all out to build up oil production now to reduce the balance of payments deficit and rely on nuclear energy to take its place, leaving the next generation to worry about the state of the economy at the end of the century? Or should we settle for conservation policies to make our reserves of fossil fuels last as long as possible, with consequent fall in growth rates, while concentrating on developing all the alternative fuels? This route would play down nuclear energy on environmental grounds, and aim at a windmill in every garden, a solar panel on every roof, a barrage in every estuary and a flotilla of 'ducks' riding the waves off our western shores. The truth lies somewhere in between the school of thought which occurring on economic growth with rapid depletion of fuel resources concentrates over the next 20 years, and the one which emphasizes the preservation of the environment without apparently wondering how we are to exist in it at anything like present living standards unless adequate energy supplies are available. The problems will not go away no matter how many placard bearing, well-meaning people confront each other in Hyde Park or, without banners, on the floor of the House of Commons. The common factor is that both on the economic and energy fronts we have to buy time. The government hopes to back the right forecast and watch the situation develop according to the appropriate scenario. The difficulty is that choice of energy policy involves commitment to capital investment and rates of growth of the economy and of depletion of fuel reserves. To have no indigenous fuel reserves at all creates one set of problems; the possession of four fuels paralyses action for fear of choosing the wrong path.

Chapter 6
The nuclear horizon

The British organization

After three decades in which we have got used to nuclear energy, it is beginning to emerge from its science fiction image as we appreciate its less glamorous but more useful applications. It is the fact that nuclear reactors release energy in the form of heat, which is used to generate steam, which in turn generates electricity, which really matters today. The growing realization that fossil fuels will probably be exhausted, except for coal, inside the next century has shifted the emphasis onto the development of safe and efficient reactor systems. Broadly speaking, nuclear power stations have a higher capital cost but much lower running costs than conventional power stations. Because of the high capital cost involved, Britain's nuclear industry has received large amounts of government money. The UK Atomic Energy Authority has responsibility for the development and promotional work required to ensure that the nuclear power the country needs is available, safe, reliable and environmentally acceptable. The development of nuclear power depends on the existence of a nuclear industry able to design and construct installations, backed by a forward-looking research and development organization. The National Nuclear Corporation (NNC), created in 1973 to replace the existing construction groups, is the single British nuclear design and construction company. It has assumed responsibility for the completion of the AGR reactors, and is active in the planning and design of the next programme of nuclear power stations. British Nuclear Fuels Ltd, in which the AEA is the sole shareholder, is responsible for the manufacture and reprocessing of fuel. The nuclear power stations, when built, are owned and operated by the electricity generating boards.

Nuclear reactors

Nuclear reactors fall into two categories – thermal reactors and fast reactors. The former have a moderator to slow down the neutrons to the energy level required for efficient production of new fissions. Fast reactors have no moderator and can utilize neutrons travelling at much greater speeds. Two types of thermal reactor power station are operating in Britain. The first is the Magnox reactor, which utilizes natural uranium metal clad in a magnesium alloy, and operates at relatively low temperatures. The advanced gas-cooled reactors (AGR) use uranium oxide fuel clad in stainless steel. Both these reactors are cooled by carbon dioxide. The prototype of the steam generating heavy water reactor (SGHWR) uses heavy water as a moderator and light (that is, ordinary) water as a coolant, producing steam which drives the turbine generator on direct cycle. The SGHWR was selected by the Department of Energy in 1974 for the next programme of nuclear power stations. Preparatory work began for the installation of four 660 MW(e) units at Sizewell in Suffolk for the CEGB, and on two similar units at Torness near Edinburgh, for the South of Scotland Electricity Board (SSEB). The research programme of the AEA was realigned to support the SGHWR decision, in conjunction with the NNC and the CEGB. This decision was reviewed by the AEA in 1976 after it became clear that the cost of the SGHWR development would be higher than anticipated and that there is not likely to be a market for this reactor abroad. In its place the AEA recommended the pressurized water moderated reactors (PWR) developed in the US. The PWR employs large pressure vessels of the type which caused difficulty in the building of the AGR power stations. It is now claimed that advances in technology have overcome the earlier problems with large pressure vessels.

Fast reactors are generally fuelled with mixed plutonium and natural or depleted uranium oxide. The AEA has a prototype fast reactor (PFR) at Dounreay in the north of Scotland. This will provide operating experience and act as a research facility for experiments associated with the production of a commercial fast reactor. The PFR has been in operation since September 1974 but the achievement of high power and continuous generation of electricity has been delayed by leaks in the steam plant. For a commercial fast reactor programme the establishment of a satisfactory safety philosophy is necessary. A key part of this is the demonstration of fully satisfactory methods of fabricating and reprocessing fuels containing plutonium. Research and development programmes relating to the treatment, storage and disposal of radioactive waste have received intensive consideration both in Britain and abroad. The whole question of the pollution of the environment, the adequacy of research on nuclear hazards and the future possibilities of danger to the environment have been the subject of an inquiry by the Royal Commission on Environmental Pollution set up in 1970.

Prospects for nuclear power

Britain, the pioneer country in the nuclear field, has slowed down nuclear development largely because of the discovery of North Sea oil and gas. In the EEC the Netherlands has, for similar reasons, postponed further nuclear construction work. The EEC countries with no significant reserves of fossil fuels have a strong incentive to build up nuclear strength rapidly, to reduce their dependence on imported OPEC oil. France and Germany have considerable construction programmes now in operation based on the light water reactors, mainly of the pressurized water type. The West German nuclear programme was suspended in March 1976 following demonstrations against the building of further reactors.

There is a prototype fast breeder in operation in France, the Phénix reactor at Marcoule in the Rhone valley, and a similar prototype is under construction as a joint German, Dutch and Belgian project. The French government is planning the construction of a full-scale fast reactor, to be known as Superphénix. It is estimated that each full-scale fast reactor will cost a minimum of £600 million with heavy associated investment in fuel plants for the processing and disposing of waste. It is unlikely that Britain could develop a system on its own in view of the vast investment already undertaken in connection with North Sea oil and gas, and the British market alone would not be big enough to support an independent nuclear industry. Experience with Magnox and the AGR thermal reactors has been disappointing in terms of exports. The US, which has successfully developed various types of light water reactor for export, is likely in the next decade to dominate the market for fast reactors. The future of the British nuclear industry depends on maintaining a capability for research and development sufficient to ensure that it plays a leading part alongside the US and Western Europe in future nuclear programmes. This means that, although Britain does not need a large contribution from nuclear energy in the short term, decisions have to be taken now over expenditure on fast reactor development. Otherwise Britain will fall still further behind other industrialized countries in the possession of nuclear technology, and in the long term will be importing not only fuel but nuclear plant and technology.

The SGHWR Case

The SGHWR programme was derived from the prototype 100 MW(e) SGHWR reactor at the AEA establishment at Winfrith in Dorset, which has been in operation for some time, proving components and fuels and gaining experience for use in commercial stations. The AEA on its present scale of operations has the resources for a realistic programme of research and development on one advanced thermal reactor system only. The choice of reactor, once made, means that effort has to be concentrated on the selected system.

The abandonment of the SGHWR programme after two years can be regarded either as a waste of valuable research time, or as an essential step towards consolidating knowledge. In many ways the recommendation to drop the SGHWR is one of the more hopeful signs for the future. For the experts to admit they were wrong is much better than Britain pouring money into what could prove in the end a dubious technological success and a commercial failure. There are precedents in other departments of experts pressing on regardless of expense.

The fact that the advisability of the SGHWR programme came under review in the autumn of 1976 underlines the advantages and dangers of the AEA position. The choice for the new programme was between the British SGHWR and the American PWR reactor. The apparent advantages of the SGHWR were that the prototype reactor already existed at Winfrith, and that when operational it would be available for sale to foreign buyers. The PWR, in contrast, was already developed for commercial use and reactors were operating in the US and Western Europe. Doubts expressed about the safety of the PWR have since been discounted. The decision to proceed with the SGHWR partly reflected the government view that the development of North Sea oil was the most urgent priority in the energy policy. The SGHWR represented an opportunity to go ahead with a British reactor which might turn out to have export possibilities, and which had the outstanding advantage of not requiring a major injection of investment capital until the 'eighties, when the prototype would have been brought up to commercial specification. If everything had gone according to plan, development work on the next generation of nuclear power stations would have gone on while the North Sea effort was at its most expensive and frantic stage. The lack of urgency in the SGHWR programme reflected the fall in demand for electricity and the fact that the CEGB's miscalculation of its forward requirements had created a considerable surplus of generating capacity. The successful start-up of the Hinkley Point 'B' AGR reactor, and the expectation that one or two other long-delayed AGR stations might be coming into action, has turned attention to the AGR as an alternative design.

Importance of the fast reactor

The case for nuclear power rests on the growth in demand for energy over the last 30 years and the fact that this has been met very largely from oil and gas, supplies of which will begin to run out in the 'nineties. The end of the world recession will, it is anticipated, bring an upturn in the demand for energy, which has been depressed since the end of 1973. Although there is no energy shortage for countries that can afford to pay high prices, a continuation of the present pattern of use will mean that little oil and gas will be left for future use. It can also be argued that the present generation of nuclear power stations can produce electricity more cheaply than their fossil-fuelled counterparts.

The weakness in the nuclear case is, first, the very high research, development and construction costs involved in bringing nuclear programmes into operation. Secondly, the time lead can be as long as 20 years from the beginning of the initial research to the time when full-scale reactors operate at full capacity. But the most critical problem is that the present generation of thermal reactors are wasteful in the use of uranium and only burn about 1 per cent of the uranium actually mined. By the year 2040, demand for uranium, assuming that no fast reactors had been built, would have reached almost 100 million tons, which is 25 times more than the latest OECD estimates of world resources of low cost uranium ore. Low cost ore is, under OECD definition, priced at $15 per pound ($26 per kg). After the low cost ores are exhausted, continuing supplies could only be obtained by exploiting progressively lower grade, higher cost ores with low energy content. These would have to be mined in very large quantities and would pose increasing environmental problems. The uranium prices in these circumstances could be as high as $100 or even $200 per pound (1975 prices).*

The shortage of low cost uranium ores means that nuclear energy faces the same problem, in a different form, as oil and gas. Assuming that the fast reactor, which can produce more than 50 times as much energy from a given weight of uranium, does not become a commercial proposition, then thermal reactors will have to rely on low grade, high cost uranium from present known sources, or hope that new discoveries will be made. Against this possibility the fast reactor appears the only solution. In the case of Britain, fast reactors can burn the 99 per cent of waste uranium that the present nuclear power stations cannot use. The 20 000 tons of waste uranium in stock in Britain could be used in fast reactors where they would produce energy equivalent to that from 20 000 million tons of coal.†

Related nuclear activities

The treatment of fuel for reactors is an important part of the nuclear industry. The preparation of fuels through the processes of purification, enrichment and fabrication does not cause problems of radioactive waste disposal. The large amounts of radioactive waste produced by power reactors are a function of their operation and of the materials of construction. The nuclear process uses up only a part of the energy in uranium and fuel can be reprocessed to separate out the important by product, plutonium, which is the fuel used in the fast reactor. British Nuclear Fuels Ltd, on behalf of the UK AEA, runs the largest reprocessing factory in the world at Windscale, where it handles more than

* UK Atomic Energy Authority, evidence submitted 1974-75 to the Royal Commission on Environmental Pollution, page 21, para 113.

† See Sir John Hill, chairman of UK AEA, 'Over-riding arguments in favour of the fast reactor', *The Times*, 29 July 1976.

2000 tons of material a year. A decision has to be taken on the investment programme for plant for reprocessing spent fuel to extract and recycle uranium and plutonium. In particular, a new 1000 ton a year capacity reprocessing plant at an estimated cost of £300 million is under discussion, to be ready by 1983. Other new plants and extensions to existing ones are under consideration. The conditions under which waste disposal is to be carried out, and the principles on which it is based are set out in the White Paper 'The control of radioactive wastes' (Cmnd 884), 1959. These various subsidiary operations are an important part of nuclear technology and represent a service which the UK AEA is uniquely able to provide for the nuclear industries of foreign countries.

Electricity demand

With the predicted future scarcity of oil and gas at economic prices, coal will be increasingly required as a feedstock for chemical and other industries. At the same time, the increased long-term demand for electricity will mean that nuclear energy will be required in increasing quantities to augment coal supplies for the power stations. Nuclear energy is, in fact, the only new and additional energy source that can be made available in sufficient quantity to meet the rapidly changing situation foreseen for the end of the century. The nuclear stations now in operation generate base load electricity at low fuel cost (see Table 6.1).

Table 6.1 Average CEGB generating costs
(new pence per kilowatt hour)

	1972-73	1973-74	1974-75
Coal fired stations	0·49	0·53	0·74
Oil fired stations	0·40	0·55	0·88
Nuclear stations	0·48	0·52	0·48

Source: CEGB.

From the table it will be seen that nuclear generating costs have remained relatively constant while increasing costs have affected oil and coal.

The 2000 AD position

The energy situation at the end of the century can be summarized as follows. Oil and natural gas will become increasingly scarce, and therefore expensive, as reserves are depleted. This leaves coal and nuclear power as the main sources of energy, helped out by any additions which may by then have been developed from alternative sources of energy, such as solar and wave power. Coal will be required for electric power stations and industrial use and as a raw material for the chemical and other industries. Nuclear energy will be used in power stations. The case for building up a large nuclear industry is that there will be no alternative source of power that can be available in time. On this basis the UK AEA has called for an independent research and development effort to be coordinated within the industry, and a forward ordering pattern for nuclear power stations with orders running at more than 1000 MW a year. This is the long-term position. In the short term, the CEGB has surplus generating capacity, the demand for energy has fallen and the Board has no immediate need to order new generating plant before 1980, and possibly 1985.

The fall in demand for electricity means that there is time to rethink the priorities of the nuclear programme. If the AGR experience, which involved creating a new type of reactor on site, has found solutions for the construction difficulties and eliminated the extra costs so far involved, then the AGR might be examined as a possible replacement for the SGHWR, or the American PWR. However, if long and costly research and development are still needed before further orders for a British reactor, whether SGHWR or AGR, could be placed, there would be advantages in moving over to the PWR, which can be factory assembled. This would, of course, mean the nuclear power industry giving up the prospect of having a British reactor to export in the 'eighties. However, given the time needed to develop a new reactor, it is doubtful whether Britain could have one available to meet the medium-term requirement for increased generating capacity.

Remember 'Concorde'

The SGHWR programme may prove to have been the last manifestation of the high ambitions built upon Britain's pioneer position in nuclear research and development. The frustrations caused by overrunning and over-spending on the AGR programme were a clear indication that all was not well with the British nuclear industry. However, the reorganizations which reduced the original nuclear engineering consortia to the single National Nuclear Corporation are assumed to have removed the main sources of difficulty.

Leaving aside political implications, the selection of the SGHWR system was backed by claims of a number of technological advantages. One was that a similar reactor is used successfully in Canada in the CANDU system. Another was that the SGHWR could be operated economically in sizes below 500 (MW(e) and so could have a large export potential. The programme was to be reviewed in 1978 and, if proceeded with, 4000 MW(e) capacity would be in operation in 1983-84. However, the SGHWR had a number of technological disadvantages. Experience showed that scaling up the 100 MW(e) prototype involved changing a number of safety features. The Winfrith prototype was designed in the 'fifties and it became clear that some basic design features would have to be altered. This explains why two years were spent preparing a 'reference design' for a reactor, the prototype of which already existed.

Past attempts to get into the reactor export business have been disappointing and there seems little point in running after a bus that we have already missed. If the nuclear industry is to cash in on previous investment it will be through its experience in related activities rather than starting out on a new British reactor programme. Reprocessing of waste, the manufacturing of nuclear components, and other aspects of the fuel cycle all demand technology which is available here. It is extremely important that Britain should maintain a strong nuclear industry to deal with the introduction and operation of the medium-term programme which may very well consist of PWR reactors operated under licence. In the background there must be a continuing body of research directed to the fast reactor for which British design has worldwide acceptance. This does not mean that Britain should attempt to bring off the fast reactor in a great solo effort. On the contrary, there is everything to be said for taking a leading part in an EEC or OECD research project, thus freeing considerable amounts of capital for investment in manufacturing industry. The first item on the agenda for the consideration of the nuclear programme should read:

Item 1. Remember 'Concorde'.

Nuclear apprehensions

The reduction of complex problems to simple statements is dangerous when these are based on unproved assumptions. The case for a massive increase in nuclear energy capacity follows on from a number of such assumptions. The progression is as follows.

1. An energy gap will occur at the end of the century when oil and natural gas supplies will be exhausted.
2. Coal will be the only fossil fuel with substantial reserves for the future. These will be needed in part for use as feedstock for the chemical industry and for making substitute natural gas.
3. Alternative sources of energy – hydro-electric schemes, wave, tidal and solar power, and so on – will only be in a position to provide a small proportion of total energy needs.
4. Nuclear power will be needed to replace oil and gas.
5. Thermal nuclear reactors burn uranium, of which there may be a shortage before the end of the century.
6. The only solution to this problem is to bring fast reactors, which produce more fuel than they burn, into operation.
7. The energy gap will be closed by coal and nuclear energy only if fast reactors are available to solve the nuclear fuel problem.

This straightforward statement appears to offer a complete solution to the problem of the energy gap. However, nuclear technology is still a comparatively new and unexplored subject, with its own problems of safety and security. Experience in the second nuclear programme, of AGR reactors, showed many engineering problems still not completely mastered. For fast reactors using plutonium a completely different set of problems arise.

The Royal Commission

The wider implications of developing the fast reactor were examined by the Royal Commission on Environmental Pollution under the chairmanship of Sir Brian Flowers. Its report,* published six years after its

* 'Nuclear power and the environment', sixth report of the Royal Commission on Environmental Pollution, HMSO, London, September 1976.

appointment, underlines the risks involved in developing a major nuclear programme. The reason for the diffidence of the Royal Commission is the fact that the fast reactor is of a radically different design from the first three generations of nuclear reactor. The prototype at Dounray has a 250 MW output and, on the basis of experience gained, the AEA is pressing for permission to build a full-size reactor. The report accepts the need for increased nuclear capacity, but argues that any decision on a major commitment should be postponed for as long as possible. It found no technical reason for opposing the CFR 1, the commerical-size fast reactor, but sidestepped the issue by stating that it was 'not for the Commission to propose or oppose development'. However, the report states quite clearly that Britain 'should not rely on something that produces hazardous substances such as plutonium unless we are convinced that there is no reasonably certain economic alternative'. It calls for a new regulatory authority to control the safety of nuclear installations and their effluents and emissions, and for a new state-owned company to be responsible for the safe disposal of all waste arising at nuclear sites. The report also states that there should be no commitment to a large programme until an acceptable solution had been found for the problem of storing highly radioactive waste.

Although the Royal Commission report underlines the dangers of expanding Britain's nuclear energy capacity, including the possibility of plutonium falling into the hands of terrorists, it does not pronounce against an expanded nuclear programme. What it does is to point out the problems and say that nothing should be done without taking full account of them. The report had a profound effect on public thinking on the energy problem. Environmentalists were able to find in it a justification for not bridging the energy gap via the nuclear route. The AEA was equally able to discover passages which acknowledged the need for an increased nuclear capacity. The lesson from the Royal Commission report, as Lord Rothschild pointed out in an article in *The Times*,* was that deciding whether or not to go ahead with the fast reactor posed terrible problems. It was not enough to brush technical and environmental problems aside and let the future look after itself. The answer must be to delay the final decision as long as possible while stepping up research on the fast reactor. At the same time there is no reason why Britain should work alone on this problem. An agreement

* 27 September 1976.

to exchange information with the US was signed at Germantown, Maryland, on 23 September 1976, the day on which the Royal Commission report was published. This coincidence would seem to indicate the future direction to be taken by nuclear research.

Nuclear fusion

The case for an extended nuclear power programme was set out in the WAES Report in terms of the three stages of choice which different countries could adopt according to their energy situation. The first of these involves a single use of uranium with used fuel stored under water, with the question of reprocessing indefinitely postponed. The second choice introduces the transport and reprocessing of fuel, which for Light Water Reactors might add fuel equivalent to roughly 20 per cent more uranium. The third choice, the fast breeder reactor, is seen as essential if nuclear power is required as a substantial source of energy. Decisions on using the fast breeder must be preceeded by the setting up of reprocessing and plutonium recycling plants. Decisions on the second and third choices are more expensive and more controversial than those affecting the operation of thermal reactors, which nowadays tend to be taken for granted. It is not easy to weigh up the risks and benefits of nuclear energy. While there are grave risks involved in moving into the fast breeder, its rejection raises problems of a different order.

In some circles great store is set on the possibility of developing a safe system of nuclear fusion, which depends on lighter elements coalescing with heavier ones, as occurs in the tremendously hot cores of the sun and stars. This is the power used in the hydrogen bomb, which is another way of saying that man has succeeded in bringing about fusion reactions without having any idea of how to contain and use the enormous energy generated. The substances used in this case are not uranium and plutonium, but the hydrogen isotopes deuterium and tritium, and the common metal lithium. The process of creating the fusion, which results in the production of helium, hydrogen and energy, involves operating in temperatures of around 100 million degrees centigrade in a condition known as a plasma.

At this point the theory comes up against a number of operations that are, in the present state of knowledge, impossible. The research into fusion has not yet reached the stage where an attempt can be made to produce all the conditions necessary for an energy-producing reaction.

Work is being concentrated at the Department of Energy research station at Culham and elsewhere on specific aspects of the problem. No one knows whether a practical system can be produced, and in the meantime large sums will have to be spent on research and experiment to find out. The fact that deuterium and lithium are freely available in seawater has led to the belief that, given a breakthrough, the world's energy problems would be solved for all time. It appears that the major limiting factor in fusion, as opposed to fission, reactors is the scarcity of rare materials such as vanadium, niobium and molybdenum needed in their construction. Because of their lower densities of energy production, fusion reactors would require a core about a hundred times as large as that of a conventional nuclear power station. On the other hand, some quite different method of construction might be devised, such as holding a fusion plasma by an immaterial container made up of very strong magnetic forces. Clearly, at this stage of knowledge it is impossible to say whether a fusion reactor could be made, and equally whether, if it was, it would be more or less dangerous than a fission reactor. What does seem to emerge from this brief survey* is that the expense involved would be far greater than Britain could contemplate alone, and that research costs in this field should be shared. The major relevant research effort in the EEC is the TORUS (JET) experiment costed at £60m over a period of five years. At the time of writing argument is going on between Italy, France, West Germany and Britain as to where it should be sited, and the future of the project is in doubt.

* For a most interesting account of fusion power see Gerald Foley, *The Energy Question*, Penguin, London, 1976.

Chapter 7
Mixed feelings about the mixed economy

When is an option open?

Projections of energy requirements must take account of possible developments in economic policy. In Britain, this involves consideration not of a free market in which supply automatically adjusts to demand, but of a mixed economy with a growing public sector. Prices in the public sector are not determined solely by demand, but also by costs, especially wages and salaries, and at the same time are controlled to meet broader economic policy objectives. In theory, at any rate, the ability of governments to supervise the prices and investment policies of nationalized industries and to set financial targets should ensure the harmonious development of the fuel industries. If special problems arise, these could be met by short-term measures, such as the paying of subsidies to meet unexpected losses, price adjustments or the writing off of deficits. Government supervision of the nationalized industries is also intended to ensure that the needs of consumers are met at reasonable prices with assured security of supplies.

So long as oil was cheap and plentiful, and under the control of the oil corporations, there was always a fall-back position for consumers. This situation has now changed. First, oil prices have risen to a level where it no longer has an advantage over coal and other fuels. Second, the formation of BNOC and the development of North Sea oil with government participation is bringing oil increasingly towards the same conditions of price and investment control as the public sector fuel industries.

Centrally planned energy policy

It is possible that the price of different fuels may become less important as a determining factor in deciding the relative levels of consumption, and that the principle of market competition will be quietly pushed into the background. This new approach to fuel pricing was set out in a document* by two political advisers to Mr Anthony Wedgwood Benn, the Energy Minister, and circulated at the National Energy Conference in June 1976. The paper argued that the main reasons for shifting over from a system of market choice to one in which fuel use is determined by government direction were that oil prices had risen, and that world reserves of oil and natural gas could be largely exhausted in a few decades. A centrally directed pricing policy would take account of the problem of how long-term energy requirements were to be met, and attempt to regulate relative depletion rates of the fossil fuels so that reserves were used to the best advantage. In short, a system of allocations with the equivalent of fuel ration books would be introduced. In the situation to the end of the present century, the authors argued, adherence to a market solution would mean that natural gas would remain relatively cheap until most of the offshore supplies were exhausted. There would be pressure to use gas in power stations and coal would remain relatively expensive while the NCB carried out its major investment plans for bringing new capacity into use. Nuclear energy would continue to be relatively dear, although the possibility of bringing new supplies onto the market through the completion of Dungeness 'B' and the other unfinished AGR power stations should not be ruled out. The emphasis in this planned policy is ostensibly on protecting gas reserves by raising the price, and pushing forward the completion of the AGR power stations rather than on starting a new nuclear programme. Although competition between fuels has been far from free in the 'sixties and 'seventies, the position has usually been one in which prices were a compromise between the attempt to cover production costs and the need to maintain a competitive position, especially in relation to oil. This is very different from the proposal in the policy paper that prices should be used to prevent one fuel having a competitive advantage, even if this were based on a favourable cost position.

* Frances Morrell and Francis Cripps, 'The case for a planned energy policy', 1976.

Market competition

Will the replacement of traditional policies of commercial competition by one of government regulation and control raise or lower the prospects of success? The first question is how successful the free market has been in providing consumers with the fuel of their choice. At the family level success has largely been a matter of luck rather than judgement. Those households that installed oil fired central heating back in the 'sixties were no longer feeling the accustomed pleasant warm glow when the bills for the winter of 1973-74 came in. Others who put in night storage heaters suddenly found themselves in the cold, not because of the change in the fuel situation but because of a change in electricity pricing policy by a hard-pressed Chancellor of the Exchequer anxious to increase revenue. The lucky ones at the end of 1973 were the families with gas central heating, gas cooking, insulated houses and small cars. Not all of them arrived at this happy position as a result of careful analysis of market prospects. Those less well placed could only improve their position by converting to gas central heating, or if they could not afford this, introducing a strict regime of switching off.

In industry many firms do not have a chance to benefit from commercial competition, owing to the restrictions on the choice of fuel imposed by technical considerations. Where the requirement was for power to drive machinery there was no effective alternative to electricity. For steam raising and processes requiring constant high temperatures, there was competition between coal and gas, with gas increasingly taking the lead where bulk supply arrangements were acceptable. Conversion from one fuel to another is costly in industry and may involve interruptions of work. For factory heating the same options as for domestic consumers broadly apply. One advantage enjoyed by the industrial over the domestic consumer is the ability to operate oil burning generators in the event of a failure or shutdown of the electricity supply. Many domestic and industrial consumers would argue, in the light of experience, that freedom of choice amounted to no more than being able to opt for a particular fuel, only to find later that the conditions on which the choice had been made had suddenly changed.

Price control is less important when markets are depressed, but it influences every assessment of potential return on new investment, especially in high risk ventures. In the economy generally, prices are the key to investment intentions. A rise in prices and recovery in profit margins is the surest way to secure new investment.

The public sector

Projections of Britain's energy requirements involve consideration of the peculiarities of a mixed economy with a fast growing public sector. The energy industries, except oil, are all publicly owned, and even in oil state participation through BNOC will introduce new constraints on prices and supplies. An element of competition exists between coal, gas, and electricity, but the extent of price changes is controlled by the Department of Energy in conjunction with the Department of Trade and the Price Code.

Nationalized industries are often criticized on the grounds that they use up excessive amounts of capital and labour, and make huge losses cushioned by write-offs and subsidies from the Treasury, instead of being required to show market rates of profit on capital invested. Certainly, the way in which some of the nationalized industries lurch from loss to profit to loss, with price increases and cuts in services coming in between, reinforces the idea that more financial discipline is required. However, to criticize public sector concerns because they do not behave like private enterprise firms is not politically realistic. Whatever the shortcomings of nationalization in general, and they are many, in considering the part to be played by the energy industries in the evolution of economic policy, nationalization must be accepted as a fact. The idea that all would be well if the NCB was subject to the same financial disciplines as private sector companies, while the latter were subsidized when necessary on the same scale as the NCB, is not simply to confuse the issue but to turn it inside out. The point about the nationalized industries is that, while few of them can pretend to be paying their way in any real sense, the volume, type, or cost of their activities affects the welfare of the population at large. Private firms in the same monopoly position as nationalized industries could very well choose to concentrate on products that gave them above-normal profits, instead of more widely used but less profitable items. The nationalized industries have a number of social obligations which are not always taken into account when profit and loss figures are presented. The problem of employment in areas where coal mining was the only industry, the use of subsidies to maintain what may be low productivity employment instead of paying out the same or more money in unemployment benefits – these are examples of the use of nationalized industries to further general economic policy aims. Similarly, welfare considerations arise at the consumer end when the costs of providing heating for the poor and elderly have to be set against the possible medical and hospital costs involved if supplies were cut off when bills are not paid. The fact that the costs of these alternatives are not compared in any official calculations is no reason for their not being taken into consideration.

Another criticism is that few of the nationalized industries are improving efficiency and that their profits are often derived from sidelines while their main activities result in losses. More seriously, they have not been able to provide for their own investment out of earnings.

Public sector pricing

The two goals of economic policy that have most effect on the nationalized industries are the levels of employment and of inflation. At different times the energy industries have been compelled by governments to employ more workers than they required. Also, because energy enters into the costs of other industries, governments have taken responsibility for the timing and size of fuel price increase. The 1970 Conservative Government used the public sector as a weapon in the counter-inflation policy by holding prices of nationalized industries down, and then subsidizing them when it was necessary to make up their revenue. This kind of action is still taken, but on a more selective basis as experience has shown that attempting to regulate the economy by manipulating prices and wages in the public sector can have unpleasant side effects.

Demand for energy is responsive to price competition only within the constraints imposed by the need to have the equipment and appliances necessary for the change from one fuel to another. The argument that the price of gas was too low in the summer of 1976 illustrates the problems that can arise in the public sector when price control and competition come into conflict. The price of North Sea gas was well below that of coal or electricity, a fact which would be expected to stimulate competition from these two industries. Instead, the response took the form of demands that the price of gas should be increased so that it would not gain too large a share of the space-heating market, for which incidentally, it is ideally suited. The gas industry found itself in this favourable position largely because the phasing of its investment programme was different from that of either coal or electricity. The major operation of converting appliances and equipment to take natural gas had been completed and supplies were coming in from the southern basin of the North Sea at relatively low cost. Continued future supplies were assured from the north basin and by purchases of Norwegian gas from the Frigg Field. Conditions, in short, were entirely in favour of gas as compared with its competitors. This convergence in time and place of a number of factors favouring the gas industry was largely fortuitous and is in any case a relatively short-term phenomenon. In the 'sixties OPEC oil was similarly favoured by economic circumstances.

The case of low priced gas introduces a further problem of energy forecasting. It is frequently argued that past mistakes have been largely due to making long-term decisions on the basis of short-term considerations. The case for higher gas prices had a measure of support on the grounds that they might prevent over-depletion of resources. If gas and electric space-heating costs were more in line with each other, more people would use electricity, which would cause more coal to be used in power stations. Thus, instead of using up natural gas, which is in relatively short supply, we could be using more coal, our most plentiful fuel. A further argument is that so long as cheap gas is available, conservation of fuel reserves and the development of renewable sources of energy, such as wind, wave, or solar energy, will not be pressed forward with the necessary urgency. In the event, after a 12 per cent price increase in October 1976, the Gas Corporation was ordered to make a further 10 per cent increase in April 1977 on the grounds that this was necessary to reduce the public sector borrowing requirement.

The NEDO analysis

In theory, at any rate, the ability of governments to supervise prices and investment in the public sector and to set financial targets should ensure the harmonious development of the fuel industries. If special problems arise, they could be met by short-term measures, such as the payment of subsidies, or help to poorer consumers by means of special supplementary allowances. The whole question of the financing of the nationalized industries has been studied by the National Economic Development Office, which published its first report on the subject in July 1976.* It concluded that the finances of the nationalized industries had deteriorated and were in a state of confusion because governments had intervened to secure short-term solutions to particular economic or industrial crises. NEDO suggested the need for a fundamental change in the structure of relationships between state industries and governments, and the laying down of financial and political guidelines within which each industry should operate. In particular, the report argued that the present separation of functions between ministries and nationalized industries should be replaced by a relationship committing all those involved to a plan for developing the public sector industries.

The NEDO proposals, if adopted, would do away with the existing system of setting targets and monitoring performance of the nationalized industries. This system, it is argued, has failed to provide the necessary disciplines to enable management to operate without day-to-day interference. Such an arrangement would go beyond the requirement of profitability on a conventional, one year with another, basis. It would be followed by the adoption of formal financial disciplines, such as the setting of target rates of return, the practice of relating prices to long-run marginal costs, and the principle of separate accounting and compensation for loss making ventures and other non-commercial objectives, which were carried on possibly at the insistence of government. In this connection there may be a case for making good the anomalies created by price restraint in the capital structures of the energy industries. Compensation paid to them for holding down prices has been limited to the deficits they actually incurred. The profits they were prevented from making by having to charge low prices have had to be made good by extra fixed interest borrowing. The result is, as the NEDO report pointed out, that the reserves of the industries are now lower than they would otherwise have been, and the burden of interest charges and debt redemption is correspondingly greater.

* 'Financial analysis of nationalised industries', NEDO, HMSO, London, 1976.

Keeping options open

The lesson of the last 20 years is that it is unwise not to keep all possible energy options open. This is easier said than done because changes in the energy pattern are costly and slow to bring about. In Britain, investment in energy supplies amounts to about £2 billion a year, or a tenth of total national investment for all purposes. Much of this money goes on installations which are different for each of the four fuels available. For coal, the requirement is for long-term investment in new pits and the improvement or extension of older ones to maintain supplies as worked-out pits go out of production. Oil and natural gas require offshore drilling rigs, pipelines, terminal installations, storage facilities, and in the case of oil, refineries and a distribution network. Options are open only so long as supplies of fuels last. The depreciation rate is therefore more important in the case of oil and gas than in the case of coal, with its much greater reserves.

The nuclear option depends on a realistic power station building programme, fuel processing plants, and proper arrangements for reprocessing spent fuel and disposing of radioactive waste. All these activities require heavy long-term investment. The maintenance of an adequate supply of fuel is a further requirement. In addition, all the energy industries have in common the need for training and recruitment programmes to maintain an adequate labour force, and a high level of investment in research and development to enable them to employ the highest level of technology appropriate to their operational requirements.

Investment and markets

One reason why the fuel industries have not achieved a high place in public esteem, is their frequent lapses in performance. A cold spell finds power stations with low coal stocks. British Rail, in the same conditions, is always surprised to find that its points freeze. Council housing without chimneys has narrowed fuel choice by producing a situation in which over a third of the nation's householders could not burn solid fuel even if they wanted to.

Decisions on the power station programme, how many stations should be built and what fuel they should use, determine the pattern of bulk demand for fuel. The CEGB can influence demand by changing the merit order of use of its power stations. This means that those with the lowest running costs are used for base load operation and the rest kept for peak load work. New coal fired and nuclear stations are generally used for base load, while oil and near obsolete coal fired stations, only come into use at peak periods. The markets for individual fuels are to some extent decided by the power station building programme which is approved by the government of the day. Fuel choice is the area where the ending of traditional policies of commercial competition would be principally felt. The temptation for bureaucrats to tidy up the loose ends of the energy situation by restricting individual fuels to specific uses is very strong. However, it would be difficult to use administrative direction to determine the entire national pattern of fuel use, unless freedom of choice were completely withdrawn. The alternative is for governments to give firmer control of development programmes and policies to the energy industries within a framework of clear guidelines, and provide opportunities for wider discussion on policy content and implementation, as proposed in the NEDO report. The need for a change in the control and financing of the nationalized industries is clear. The question is whether governments will be able to take the plunge and give them greater independence in their management and accountability, necessitating some new kind of institutionalized arrangement.

Policy guidelines

What sort of guidelines should be given to any organization set up to coordinate the policies of the energy industries? To be successful an energy policy must do four things. First, it must make itself acceptable to the fuel industries, to their employees, and the consumers. A policy which requires an industry first to shed workers and then to try to get them back again and raise output, as happened to coal in the 'sixties, is bound to lead to distrust and confusion. Second, an energy policy must manage to get within striking distance of matching total demand to total supply, and getting the mix of demand and supply right. This means there must be enough of each fuel to meet the specific demands for it. A total supply made up largely of coal will not keep the airlines flying or transport moving. Also, if a large part of the population is equipped with gas central heating appliances, it is a bit high-handed to say 'let them burn coal'. The totals and the mix are equally important.

Third, an energy policy must not take up too large a share of the available investment capital. There are those who believe that too much money has been sunk in North Sea oil to the detriment of UK manufacturing industry, much of which badly needs re-equipping. There is no doubt that the ill-fated AGR programme, which has so far cost many times as much as was budgeted for, has taken an undue share of scarce capital resources. Fourth, an energy policy should take account of new technologies which may require fundamental changes in the case of particular fuels. It is very important that the nature and extent of such changes should be fully explained to the workers in the industries concerned, and the measures for dealing with their impact should be a matter of discussion and arrangement through the joint consultation and negotiating machinery available. The fuel industries are an essential part of the economy, and if they are not efficiently managed and operated, manufacturing industry and the community as a whole suffer. Power cuts and three-day weeks are the inevitable result of mistakes in running the fuel industries. Above all, it must be remembered that reserves of coal, North Sea oil and gas are not only of value to the balance of payments, but to the balance of the economy.

Fuel efficiency

Another factor of considerable importance in assessing the use of different fuels is the question of relative efficiency. In the fuel industries considerable losses of power occur at the different stages of production. The overall thermal efficiency of generation of fossil fuels is only about 35 per cent. Transmission and distribution losses further reduce the thermal conversion efficiency to about 30 per cent. The transmission or transporting of energy is also a critical factor. For conveying similar unit quantities of energy over a fixed distance, comparative cost ratios are:

For an oil pipeline	1
For a gas pipeline	2·5
For a coal train	5·5
For a 500 kV electrical transmission line	17

It is clear that oil and high pressure gas pipelines have considerable cost advantages. The level of investment costs for particular fuels does not affect their thermal efficiency. It does raise the question of whether the highest or lowest cost fuels in terms of price and investment should be used up first or saved for later. The optimum solution would be to go for low cost resources first and work up to the higher cost fuels later. If the most costly resources are exploited before the cheaper ones, investment, labour and materials would be directed into energy production sooner than they need be. In other words, consumer demand would have to be directed through energy policy to use fuel reserves in the order selected, so that the lowest cost fuels were used up first, then the next least costly, and so on. In Britain, with four fuels available, this would mean working up from natural gas to oil to coal and nuclear energy.

In practice this orderly solution is not possible, as free trade in energy products does not exist. Prices are not based on real costs but reflect various non-commerical factors as well. And, most important of all, in conditions of scarcity governments do not want to be dependent on imports which can be interrupted for political reasons. Each country tends to use the fuel that is most readily available with reasonable security of supplies, whether or not it is the cheapest. The basic policy priorities therefore come down to economy in the use of energy, encouragement of production of indigenous supplies, and securing imports, as far as possible, from countries that give the greatest assurance of continuity of supplies.

The vicious costs circle

While the economy cannot function without energy, the amounts required depend on the level of economic activity. Part of the energy used is needed anyhow, whether business is booming or depressed. Houses and streets must be lit, fuel provided for domestic use, and for road, rail and air transport. If the economy is booming, production and employment will increase and companies will invest in new plant and capacity. The demand for energy in industry will rise, which will mean using more coal, oil and nuclear energy to generate electricity, more oil for transport, and more gas for industrial process work and heating. The opposite effect was seen in the recession 1974-75. World oil consumption fell in both years, and in 1975 world oil production fell for the first time since 1942, the main reduction occuring in OPEC states and in Latin America. Excluding the USSR, Eastern Europe and China, the production of oil fell by 8·5 per cent as a result of the recession and the running down of oil stocks.

The problem of long lead time between investment decisions and new plant coming into use is relevant even when business is slack. New refinery capacity came into operation in 1975 and the building of new refineries and power stations went on throughout that year. In spite of scrapping many tankers and converting a large number of others into dry or special cargo carriers, the average tanker surplus rose in 1975 to over 110 million tons deadweight. Nevertheless, some 88·5 million tons deadweight of new tankers were on order at the beginning of 1976.

This reaction to the 1973 oil price increases gives some indication of the difficulties involved in forecasting. Although consumption of oil fell, it is not really possible, so far as the industrialized countries are concerned, to say how far this was due to price increases and how far to the world recession. The fact that long-term capital projects went ahead regardless of the slow-down in business, shows that capital and consumer goods tend to follow separate trends in the short term. The major constraint on oil consumption is likely to be the level of production set by major producers. If Saudi Arabia, which does not need to raise the level of its oil revenues, decided to cap its production at the present level of 9 million barrels per day, then world oil supply would fail to meet demand inside four years.

Coal, which could be used as feedstock for both oil and gas production, could prolong the use of both, but the processes involved are expensive. A vicious costs circle comes in here as capital investment on coal conversion plants could divert capital from manufacturing industry. While there is plenty of coal in Britain for the next century and beyond, the rest of the world has to take account of the concentration of resources in the USSR, China and the US. Uranium supplies are already giving rise to concern and even with no economic growth at all we could not look to nuclear power for a solution to the energy shortage unless the fast reactor could be brought into operation.

Yet another aspect of the vicious costs circle is that without heavy expenditure on research and development there is no hope of solving the technical problems of nuclear fusion, or of the various renewable sources of energy. In other words, we may not be able to raise economic activity to the level where it will support investment on long-term energy requirements and at the same time provide the wherewithal for the capital investment needed to raise industrial production. The British economy is too debilitated as a result of inflation and failure to modernize its political and industrial structure to mount a decisive research programme of its own. The best solution would be for a large international development programme to be agreed on within the framework of the IEA or the EEC, in which Britain could help in determining priorities and working for the solution of the problems they represent. One form such a programme might take is outlined in Chapter 8.

Chapter 8
Would a common energy policy help?

Common policy initiatives

In the days of the six-nation EEC, movement towards a common energy policy was barely perceptible. One reason for this was the separation of responsibilities between the EEC (oil), European Coal and Steel Community (coal), and Euratom (nuclear energy). The merging of the three communities and the creation of a single executive in 1967 brought the major fuels under the responsibility of the new Commission. In the energy situation as it then was, the main concern was the rapid run-down of the coal industries of the Six and the extent to which the social costs involved should be met from national or EEC funds. Plentiful supplies of cheap oil from the Middle East and North Africa, and the availability of natural gas in the Netherlands, France and Italy more than compensated for the decline of coal in the energy balance and the very slow progress in the build-up of nuclear energy.

In December 1968 the Commission produced a memorandum entitled 'First guidelines for a Community energy policy'. This document was concerned with what might be called the market aspects of Community philosophy. It argued that energy policy should be based on the interests of the consumer, because increases in energy prices affected the competitiveness of industry and therefore the cost of living. The thinking of the Commission was on broadly the same lines as the 1967 White Paper on fuel policy (Cmnd 3438) which introduced the four-fuel policy in Britain, with the emphasis on competition and freedom of choice between the different fuels as a means of ensuring stability, security and cheapness of supplies. Within the EEC, distortions were to be removed by the free movement of supplies and the elimination of impediments, whether due to the activities of governments or to technical obstacles. In particular, the memorandum stressed the need for the harmonization of taxes in the energy sector. Community help in the reorganization of the coal industry was called for as well as the better coordination of national aids. Security of supplies of oil was to be ensured by each member government maintaining stocks equal to 65 days consumption.

Steps by the Six

After the publication of this pioneer memorandum, but before the enlargement of the EEC, the Six took a number of steps along the road towards a common energy policy. Regulations were introduced approving a general plan of aid to the oil industry, and to producers of coking coal and coke. Regulations issued on 18 May 1972 requiring information to be given to the Commission on imports of hydrocarbons and an investment programme for oil, natural gas, and electricity could be regarded as the most important steps taken by the Six towards the setting up of the machinery of a common energy policy.

In January 1973 Britain, Denmark and Ireland joined the EEC and subsequent discussions highlighted the conflict between national and EEC interest. Did Britain as the producer of over half the EEC's coal, with newly discovered reserves in East Yorkshire and elsewhere awaiting exploitation, and with as yet unquantified resources of oil and natural gas in the North Sea, stand to lose or gain from participating in a common energy policy? An alternative but scarcely less controversial formulation of the same question was to ask what kind of common energy policy it would be in Britain's interest to accept, and how far this would be appropriate to the economic objectives of the enlarged EEC.

Nine energy policies

The energy policies of the EEC followed the general pattern of cheap and abundant supplies of imported oil replacing indigenous coal, where this existed. Today, only Britain and West Germany still have important coal industries. The Dutch have closed their last remaining mines, while French output has fallen to below 30 million tons a year and Belgian output to around 10 million tons. British production, at 125 million tons, is equal to that of the rest of the EEC put together. In the main, EEC coal production is deep-mined and relatively high cost and, as in Britain, the principal consumers are the electricity generating authorities and the steel industry. The coal industry is subsidized in all the producing countries, the highest rates being paid in France and Belgium. Imports come from Poland, Australia and South Africa, and conflict could arise over the extent to which these might displace Britain's exports to the EEC.

The major obstacle to securing agreement between the different states on a common energy policy has been the existence of differing interests and attitudes. Those without coal have seen no virtue in supporting the coal industry in other parts of the EEC. Those housing the headquarters of major oil corporations, notably the Netherlands, take a different view of oil supply and pricing policies from the rest. Italy has struggled to build up an oil industry with supplies from a wide variety of sources. France, on the other hand, has concentrated on oil from Algeria, with pricing policies to fit membership of the franc zone. Superimposed on the differences in national energy resources is the divergence in attitudes between West Germany, with its preference for *laissez faire* market policies, and the French with their *dirigiste* tradition.

National interests come first

The area of energy policy where attempts at cooperation between the Six met with most difficulties was in Euratom. Here the individual states refused to compromise their own nuclear interests in order to operate a Community policy. France and West Germany developed nuclear programmes of their own and the Euratom common research effort was largely frustrated. While there is broad agreement now on the need to undertake common research policies, commitments to specific technologies have probably gone too far for any meaningful co-operation in the immediate future. In the North Sea the four member countries which have resources of oil or natural gas – Britain, the Netherlands, Denmark and West Germany – are not disposed to see these pooled for the general advantage. All this means that the conditions for setting up a common energy policy are far from favourable. The emphasis has always been, and continues to be, on national rather than Community interests.

The Commission has taken the view that competition must provide the main stimulus in the energy sector. While it is agreed that governments must have the power to intervene, there is room for considerable differences of view on when intervention should take place. The Commission is in no position to direct the actions of member governments and in its statements has opted for a supervisory role. There is still a long way to go, however, before the job specification for this role can be drawn up. The greatest enthusiasm for a common energy policy is to be found among the bureaucrats in the Berlaymont. Their counterparts in the capital cities are less hopeful of success.

Policy after the oil crisis

When the energy crisis broke in October 1973 the EEC was still without a common energy policy. In May 1974 the Commission submitted a Communication to the Council entitled 'Towards a new energy policy strategy for the Community'. This was followed by various proposals for individual fuels which together formed a complete and coherent programme for a future Community energy policy. The main object was to secure adequate supplies at prices which did not prejudice economic development. To do this a coordinated energy policy was proposed with the following aims:

1. To increase substantially the production of nuclear electricity.
2. To maintain the production of coal at current levels to 1985 and beyond, with a small increase in imported coal supplies.
3. To stabilize the use and consumption of crude oil and so reduce oil imports.
4. To increase EEC supplies of gas.

The Commission estimated EEC requirements for 1985 at 1475 million tonnes of coal equivalent of which solid fuels would account for 17 per cent, oil 41 per cent, gas 23 per cent, nuclear energy 16 per cent. It was emphasized that these figures were only proposals and not forecasts and that the amounts would be revised from time to time. Furthermore, as the objectives were global and gave no precise amounts for different member countries, the proposals had the unreal air of an academic exercise. However, the broad objectives for 1985 were to reduce the total use of energy by conservation and to expand electricity consumption without increasing dependence on oil. Dependence on imported energy supplies was to be reduced from 63 per cent in 1973 to 45 per cent in 1985, by which time nuclear energy should cover 50 per cent of electricity production, giving a total nuclear power station capacity of about 200 GWe, compared with 11 GWe in 1974. Internal production of solid fuels including lignite and peat should be maintained, although the anticipated annual loss of coal capacity would require considerable investment to replace. Imports of coal would be needed above a level of 300 million tonnes production. A doubling in the internal production of natural gas, together with a substantial increase in imports of liquefied natural gas, was regarded as essential. Consumption of crude oil was expected to reach its peak in 1980 and fall to its 1973 level by about 1985. This would be achieved by increased use of indigenous supplies, mainly from the North Sea, reducing total energy consumption by concentrating oil consumption on specific uses. A small increase in the use of hydro-electric and geothermal energy was forecast. Finally, the proposals called for the allocation of research funds to discover new sources of energy.

The 1974 proposals also stated that financial guarantees and assistance would be made available to projects in member states which were in line with the long-term objectives of a common energy policy. The British coal industry, for example, benefits from financial help and guarantees to enable it to reach the output targets for 1985 set by the EEC and the British government. Similarly, the expansion of the British nuclear energy programme attracts increased financial help from EEC sources. EEC finance is also available for the development of North Sea oil, reducing the heavy financial dependence on the international capital market. The Commission emphasized that provision of EEC finance for North Sea oil production did not imply any interference with Britain's sovereign right to control these resources. It went on to say that member states had recognized treaty rights over the exploitation of Continental Shelf resources off their shores. They were free to decide the rate of exploitation of oil and gas, and to nationalize any economic activity if they so wished. The only concern of the EEC in relation to North Sea oil was to ensure that it was offered for sale to consumers in other EEC countries without discrimination in terms of price or of quantity. Britain is highly sensitive about the development of North Sea oil and has made drilling licences conditional on all oil being landed in Britain except under a special waiver from the Energy Minister.

Is Britain right?

The need for a common energy policy only arose in its present urgent form after the 1973 oil crisis. Before then, although indigenous supplies of fuel were just as unevenly distributed as now between the Nine, cheap OPEC oil was always available. Increased oil prices and the prospect of an interruption of supplies have coincided with the development of North Sea resources. Although there are certain similarities in the energy situations of the member countries, national differences make it virtually impossible to devise a policy that would suit all of them equally well. This explains why, although the example of the common agricultural policy (CAP) is often cited, the Nine have not yet been able to agree to act together on energy. However, the principle on which the CAP operates could be applied to the energy field. Broadly, the CAP guarantees markets for an official list of agricultural products and uses various fiscal and other devices to boost their production to the level of EEC requirements. Foreign supplies are excluded where a sufficient degree of self-sufficiency has been achieved, and exports of surplus products are subsidized. The disadvantages of this system are that prices to the consumer are often above world prices, and that mistakes in the level of support to producers lead to aberrations such as butter mountains, wine lakes and the margarine price controversy. However, in spite of difficulties incidental to its method of operation, the CAP has for the most part provided security of food supplies at EEC rather than national or world prices.

The fundamental question regarding any possible common energy policy is: who is to control indigenous resources, national governments, or EEC bodies working under the control of the Commission? The common agricultural policy is controlled from Brussels, with results that are not regarded as satisfactory by all member countries. The same mistake need not be made in relation to energy policy.

Prospects for agreement

The possibilities of agreement on a common energy policy are not great, no noticeable advance in this direction having been made since the summit meeting of October 1972 which called for the formulation of an energy policy 'guaranteeing certain and lasting supplies under satisfactory economic conditions'. This is the kind of generalization that is acceptable in a communiqué but difficult to translate into a programme for action, as there is no way of guaranteeing supplies, even in a country like the US with vast indigenous resources. The most secure fuel may also be the most expensive, the cheapest may have the highest pollution level or the highest environmental cost. When it comes to reconciling national policy objectives, the starting point is the level of indigenous resources. Energy-importing countries are more enthusiastic for a pooling of resources than those with indigenous supplies. Countries with important coal industries are more interested in the optimum use of coal than those whose coal mines have been closed down.

Britain and the common policy

From what has happened in other sectors where common policies have been introduced, it is clear that countries which go into negotiations with firm policy objectives are most likely to come out with policies which suit their interests. The failure of the Six and the even greater difficulty of the Nine in reaching a common energy policy had advantages for Britain in the early years of its EEC membership. With the greatest reserves and the highest current production of coal in the EEC, together with unquantified resources of oil and gas in the North Sea, Britain has a considerable interest in the form of the common energy policy. In the present weak state of the economy, the temptation is to regard North Sea oil as the ace in the hole which must be played for all it is worth. However, membership of the EEC covers the whole range of economic, social and monetary problems, so that unwillingness to cooperate on energy policy could have an adverse effect in other policy areas where the advantage lies with other EEC members. It will therefore be to everyone's advantage if a way can be found to reconcile national with EEC interests. As production of oil increases the need for agreement will become increasingly important.

Community energy balance

The impact of North Sea oil on the EEC energy balance is more than a matter of how the statistics are presented. Ideally, oil and gas from the

North Sea should be regarded as a welcome extension of the EEC's energy base, providing it with a countervailing influence in the world economy of the 'eighties, which at present is sadly lacking. The question is whether the British government regards EEC membership as sufficiently important to warrant joining in a common energy policy requiring some measure of agreement over the disposal of North Sea oil production. The answer will be determined by the extent to which a common energy policy affects revenue and foreign earnings from North Sea oil, and therefore the efforts of the government to break free from its on-going economic problems.

The advantages of concentrating trade in North Sea oil within the EEC would be considerable for Britain and the Eight. Far from being able to take a leading place in EEC affairs, Britain has, despite some skilful window-dressing, presented a very low profile. Production of North Sea oil has already reduced the general balance of payments deficit but will not solve the long-standing weaknesses of the economy. Some of the earnings from oil sales may find their way into industrial investment, and there will be some improvement in this direction as the flow of investment into North Sea development eases, leaving much needed funds available for use in other sectors of the economy. Either way, new industrial investment is necessary to put right some of the chronic structural faults in the economy. However, there are few absolutes in economic policy. An improved balance of payments due to expanding oil revenue may cause a rush of confidence to the heads of foreign holders of sterling, push up the exchange rate and make our exports so costly that the oil revenue will have to be used to rescue a rising pound.

What kind of policy?

The search for an acceptable common energy policy is complicated by the differences in the time scales of investment and market decisions. Changes in investment policy take a considerable time to come into operation. At nine-nation level, policy makers seeking to reconcile divergent objectives based on different resource positions, geographical situations and industrial structures are faced with a promethean task. Any EEC energy policy must be worked out on a highest-common-factor basis. The only component of the energy equation common to all EEC members is that they are all oil importers, and apart from Britain, which should achieve self-sufficiency in the early 'eighties, they will continue to be so. Even Britain will have to import some heavy Middle East crudes to mix with its light North Sea crudes. The second common factor is that all of the Nine are trying to reduce their dependence on imported oil by increasing their use of other fuels. The EEC Energy Ministers set a target of 50 to 60 per cent energy independence by 1985, at their meeting in December 1974. This involves using more coal or natural gas, and expanding nuclear energy programmes.

The role of coal

Imports of coking coal have previously come from the US, but the energy crisis increased internal demand with consequent cuts in exports. In any case, rising production and transport costs led to higher prices for American coking coal and this upward trend is likely to continue. The other supplier of coking coal to the EEC steel industry is Poland. Here the possibility of expanding production is limited and prices are likely to follow the upward movement in world prices, unless the Polish government has an urgent need for foreign currency. It is unlikely that supplies could be obtained from producers farther afield, such as Australia and western Canada. One item in a common energy policy must therefore be provision for the expansion of production of coking coal in the EEC to offset the higher cost of imports. This will involve measures for the encouragement of research and development on the use of different types and qualities of coal for coking purposes, the blending of indigenous with imported coals, and the improvement of blast furnace technology to economize on the use of coke.

Fuel for use in power stations represents more than a quarter of total EEC energy requirements, with coal closely followed by oil. In the Nine, the power stations will require an additional 200 million tonnes of coal equivalent by 1985, and this could be more in the event of any shortfall in nuclear programmes. Since the 1973 increases in oil prices, coal is now cheaper than oil for power stations, in some cases by a substantial margin. The electricity supply industry in the EEC is dependent to a considerable extent on imports of OPEC oil and Polish coal. For the four EEC countries with indigenous coal supplies, the obvious policy is to substitute coal for oil in the power stations provided cost margins are not too unfavourable. The NCB plans to increase cross-Channel shipments of power station coal in the years ahead.

In the more distant future, the fact that coal reserves will last longer than those of oil or natural gas means that its use will continue and its importance increase well into the twenty-first century. There is a strong case for believing that as other fossil fuels become exhausted coal will be used increasingly for the production of substitute natural gas and as a feedstock for the chemical industry. As coal replaces oil as the basic fossil fuel after the year 2000, it will again become the focus for the energy policies of the industrialized countries, but the manner of its use will be very different by then.

EEC policy for coal

The future of the coal industry in the EEC is centred round the policy for coal formulated by the Council of Ministers. This provides a package of measures to safeguard markets and underpin national policies by EEC aid. These give financial aid for supplies of coal to the electricity and steel industries, and provide for stocks to be financed during periods of temporary fall in demand. In addition, import safeguards can be introduced to avoid the displacement of coal supplied by Britain to other EEC producers by imports from other countries. The EEC production objective over the next 10 years is 250 million tonnes a year. Within the EEC the movement of coal is likely to be from Britain to France and West Germany, and there are good prospects for additional exports from British ports provided supplies are available. Financial aid from the EEC to the British coal industry in grants for research and development, redundancy payments, training allowances and other benefits, loans for colliery investment projects and housing modernization, exceeds the annual payment of about £2·5 m by the National Coal Board to EEC funds.

The expansion of the British coal industry is based on the NCB's *Plan for Coal*, published in June 1974 and endorsed by the Coal Industry Examination, following three-cornered discussion between the government, the NCB and the trade unions. Perhaps the most important aspect of the plan in the long term, is that it represents a resumption of investment in the provision of new capacity. The proposed expansion represents a welcome reversal of the downward trend of the 'sixties. Whether Britain will be able to increase exports to the other EEC members depends on the success of the proposals for increased production. There are no obvious reasons why Britain should not step up the present two million tonnes a year exports in the next few years. Even if it does, the EEC countries will still need to import coal in the long term to offset reductions in oil supplies. None of the likely suppliers have abundant surpluses available for export so that increased financial help from EEC sources to raise the level of British exports to the other member states would be fully justified in present conditions.

The policy for coal is, so far, the only part of the common energy policy on which a measure of agreement has been reached. As such, it would be true to say that provided what has been agreed is in fact carried out, coal has a clearly defined place in the EEC policy.

The cold North Sea

The big question mark over the common energy policy concerns the development and use of North Sea oil. At present, only Britain has located major oil fields in its sector of the North Sea and brought oil ashore. West Germany and Denmark both have sectors for development adjoining their North Sea coastlines, but so far little has been achieved in the way of exploration and development. There are no indications that either country will make a major contribution to the offshore oil resources of the EEC. Holland has already discovered natural gas in its sector of the North Sea, and this will be valuable in supplementing the vast reserves in the Groningen field. Paradoxically, the largest disposable surplus of North Sea oil is in the Norwegian sector, which is one of the reasons why the Norwegians voted against EEC membership. Oil exports from Norway are likely to go to other EFTA countries as well as to the EEC. They will not displace a significant volume of OPEC imports, however, as the Norwegian government has decided on a modest rate of depletion of its oil resources.

Basis of agreement

The possibilities of friction arising from the discoveries of North Sea oil are more complicated than would at first appear. Within the EEC they concern the relationship between Britain and the other eight members. Elsewhere problems exist or may arise, between Norway and the USSR, between NATO and the Warsaw Pact countries, and between

North Sea oil producing states and countries outside Europe. If a 'have' and 'have not' situation develops in Western Europe over the possession of North Sea oil the discoveries might prove to have long-term adverse effects. A divided EEC would be badly placed to withstand pressures from the USSR, or to increase its influence in world affairs. By contrast, agreement on the use and development of the oil resources would give the EEC a new strength based on increased independence of outside oil supplies. This would involve securing EEC agreement on the price of oil; cooperation in raising funds for investment in its development; and an agreed rate of depletion combining the reduction of OPEC imports with reasonable conservation of reserves. The financing of reserves and stocks of oil would be a key issue in reaching agreement on a common energy policy.

What appears likely to happen to oil supplies can be looked at in two ways. The optimistic view is to forecast British oil production building up to between 100 million tonnes and 130 million tonnes by 1980 which, given further increases in natural gas supplies and increased use of coal, would mean complete independence of imported oil, except for imports of heavy crude needed to make up the refinery balance. As a result of the recession and the economy measures taken, oil consumption in 1980 is expected to be at about the same level as 1974, after allowing for a modest increase in the economic growth rate. In this optimistic view, Britain would be in a position to begin exporting oil early in the 'eighties. This would help to pay for the investment costs of the North Sea oil, service loans secured on the strength of the oil reserves, and help reduce both the oil and non-oil balance of payments deficits.

The pessimistic view is that the effects of the slowing down in activity in the North Sea during 1974-75 due to uncertainties about tax and participation policies, the growing realization of the unexpectedly high production costs, and difficult technical problems may prove to be permanent. The political problems need to be considered, raised by the claims of Scottish Nationalists to exploitation rights over oil fields off the coast of Scotland. The prospect that production might be disrupted by major accidents has to be taken into account. There is also concern that the rate of exploitation will be too rapid, for balance of payments reasons.

The critical choice

British policy objectives, whether secured through national or EEC policy, involve the balancing of long- and short-term considerations. While there is no doubt that the possession of the EEC's major reserves of oil and coal is of immense importance to Britain, policy decisions must be taken within the context of the world energy situation. Possessing only 1·25 per cent of total oil reserves outside the Communist world, Britain is in no position to make any significant impact on international markets but could make bilateral deals with individual countries. What the British government has to fear most is a dramatic fall in oil prices which would upset the present (to it) happy relationship between North Sea production costs and OPEC asking prices.

A common energy policy

The indications are that a 'go-it-alone', all-British energy policy would be yet another example of long-term decisions taken in the hope of short-term advantage. In the immediate short-term it would not be realistic to attempt to set up an EEC common energy policy that covered all fuels. Coal presents no problem and the implementation of the NCB *Plan for Coal* would fit into the EEC coal policy without difficulty. Nuclear energy policies are running on parallel lines and differences in systems between EEC countries are not important, provided joint policy arrangements exist for the purchase and storing of uranium and for research on the development of the fast breeder reactor, and on the so-called alternative energy sources, wave, wind and tidal power. It is only on the question of North Sea oil that policy differences appear to be in conflict.

The way to a solution would seem to be to break down the processes of oil production and marketing in terms of activities, and timing. Major oil bearing areas below the sixty-second parallel have now been discovered and recoverable reserves charted and estimated. Oil is coming ashore at a rate which should make Britain self-sufficient during 1980. The various ancillary activities, such as the construction of platforms, have built up adequate and, for some purposes, surplus capacity. What has not been decided is the rate at which resources will be depleted to 1982, and the policy regarding the size and destination of exports. Government policy appears to favour selling as much oil as possible on international markets, including the EEC in order to move the balance of payments out of deficit without depleting oil reserves too rapidly. The danger is that a balance of payments oriented policy could be self-defeating as once the more easily produced resources have been exhausted the remaining reserves would be in small, high cost fields.

EEC reserves and stock

The alternative would be to look at oil revenues over a longer period and set a value not only on oil produced but on oil left in the ground, as was done by the US government in postponing development of the Prudhoe Bay field in Alaska. Untouched reserves will inevitably gain in value over the years. If these are regarded as the reserves of the EEC, ready to be exploited after a certain date, their value could be determined in relation to current prices, anticipated demand and possible inflation rates. It should not be beyond the ingenuity of the Commission, or the Department of Energy, to work out the principles on which a part of the North Sea oil could be designated as EEC reserves for which payment would be made to the British government over an agreed period. This payment would be in part an advance on the price to be paid on the eventual delivery of the oil to the governments of the Nine, and partly a payment to compensate the British government for forgoing the return from present sales. Its importance lies in the fact that EEC payments for reserves left intact would help reduce both the balance of payments deficit and the pressure to deplete the developed fields too rapidly. At the same time, arrangements could be made for the creation of a restricted EEC reserve stock under control of the Commission as a guarantee against the interruption of supplies. The cost of creating and servicing this stock would be borne by EEC funds.

The details of such a scheme would require careful assessment. Its basic idea is conservation of supplies with provision for compensating the British government for deferring the exploitation of North Sea reserves. In its most elaborate form it could link North Sea oil reserves to a Community scheme for solving the structural problems of the British economy.

The political advantages of such an arrangement to Britain and the EEC would be considerable. The Nine backed by known oil reserves would be in a much stronger bargaining position in the world generally.

Their situation in terms of resources would be comparable, but on a reduced scale, to those of the USA and USSR. Agreement within the EEC over the disposal of North Sea oil would remove a source of friction and make it possible to move forward with other forms of economic cooperation at present hanging fire. For Britain, it would mean the possibility of attaining medium-term economic objectives without exhausting the major oil reserves. It would also provide help with the costly re-equipment of British oil refineries to deal with light North Sea crudes instead of customary heavy Middle East crudes, which Britain would otherwise have to bear alone. The sale of oil to the Eight rather than to, say, the US should not present any major difficulties and could be subject to individual contracts.

The treatment of North Sea oil in the negotiation of a common energy policy is likely to be a significant test of the strength of Britain's commitment to the EEC. However, without some common policy of this kind, Britain is in danger of losing the advantage of the biggest piece of good fortune to come her way since the industrial revolution. Furthermore, on the basis of past forecasts of what is likely to happen in the energy field, there is everything to be said for putting the emphasis on conserving rather than using up the resources we have.

Chapter 9
Looking ahead – not all gloom

Faulty forecasts

The only point on which there is any measure of agreement on past energy forecasts is that they were generally wrong. This is not altogether surprising, as forecasting involves making allowances for a range of increasingly independent variables affecting home and overseas markets. Somewhat unfairly, the typical forecast is regarded as an extrapolation on all demand trends set against an assumption of no increase in supplies, ending with warning bells and a retreat to panic stations. Certainly the evidence of past White Papers does not provide any convincing proof that energy policy making is more likely to be accurate if entrusted to governments than if left to market forces. Failures have not been due so much to inadequate forecasting as to the tendency for governments to assume that what they propose will in fact come about. The 1967 fuel policy White Paper assumed that cheap oil would continue to flow and that the AGR reactors would come on stream according to schedule, and produce electricity at very low cost. At a later stage, although many voices were raised to predict rising oil prices in the Middle East, no one managed to foretell the October price revolution of 1973 in all its stark reality.

Constituents of energy policy

Before looking at any of the available forecasts it is important to be clear about the factors involved in arriving at a coherent policy. Britain now has huge reserves of coal, estimated to be sufficient for 300 years, North Sea oil and natural gas which are variously expected to last for between 20 and 40 years, and nuclear technology which will be producing increasing quantities of electricity as the end of the century approaches. This would appear to be an ideal situation. If the supply of one fuel falls off, there are three others to bring forward.

The choice of energy policy is complicated by two special characteristics. The first is that commercial, economic and technical conditions are different for each fuel. In the second half of the 'seventies conditions are favourable to natural gas, which is readily available at low production costs. Conditions for coal and nuclear energy will improve as new investment decisions take effect and various market changes take place. But this will take 15 to 20 years. North Sea oil, by contrast, is now beginning to overcome some of its earlier technical problems and production is increasing steadily. At some time in the early 'eighties conditions will really begin to favour North Sea oil.

Investment's long leads

This difference in phasing is due to a number of factors but principally to the long lead times involved before investment decisions come into operation. Building a new coal fired power station can take 10 years, a nuclear power station up to 12 years, sinking a new pit 10 years. To bring an offshore oil well into production and get the oil ashore can take 5 or more years from the start of exploration. In market terms this means that rapid increases in supply are difficult to bring about. It is not possible to call for big increases in coal or oil or gas output in the short term. All through the 'sixties Lord Robens, and his successor as Chairman of the National Coal Board, Sir Derek Ezra, warned against the too rapid run-down of the coal industry on the grounds that once pits were closed they could not be opened up again. In the event, imports of oil rose rapidly while coal output dropped well below the 200 million tons target which Lord Robens regarded as the margin for national safety.

The broad options

These two characteristics of the energy industries, the phasing of the optimum conditions for individual fuels and the long leads in investment times, influence the whole range of policy decisions. In the British case they are the main considerations in determining which of two broad policy options governments should follow for the rest of the century. While there is general agreement that conditions are generally favourable the question is how long will they last, and how should the four fuels available to Britain be exploited?

The first broad view of the situation is that British self-sufficiency is not likely to last much beyond 1990. Unless action is taken now, Britain will be facing rapidly increasing import requirements at a time when energy, on world markets, will be both scarce and dear. The second view is more expansive and assumes that further discoveries of North Sea oil and gas will prolong the period of self-sufficiency well beyond the year 2000. It is also argued that ample coal reserves will still be available and that by then alternatives to the fossil fuels in the form of nuclear fusion or one or other of the various renewable energy sources may have come into operation. On this view the level of imports required after the exhaustion of North Sea oil and gas would be easily within our capacity to support.

These assumptions reflect different ways of assessing the development of the energy situation over the next quarter century. At this point no one can say precisely when North Sea oil will run out. It is more likely that improvements in technology will make it possible to take more oil from existing fields, or will make the smaller and more remote fields viable, than that some vast new field will be found. The prospects for the Celtic Sea and Atlantic Approaches have not been rated highly, but it must not be forgotten that the original prospectors worked for seven years in Saudi Arabia before striking oil.

The nuclear view

Yet another view follows on from the assumption that North Sea oil and gas will be finished by the 'nineties, but raises difficulties over the availability of nuclear energy and coal to take their place. This analysis of the situation was advanced by Dr Walter Marshall, former Chief Scientist at the Department of Energy, who argued that by the end of the century coal will be too valuable to burn in power stations, as it will be needed as the feedstock for the chemical industry and for making substitute natural gas. This means that we shall be looking to nuclear energy to generate the greater part of the electricity consumed. If this view is correct the development of the fast breeder reactor becomes a matter of urgency, as it is the only known way of overcoming the shortage of uranium which threatens to limit the total capacity for nuclear power stations. According to Dr Marshall, if fast breeder reactors, which produce more fuel than they burn, are not ready soon after the end of the century, there will be a world shortage of energy after the year 2000. The high radiation hazards of fast breeders are an additional reason for stepping up research and development on methods of dealing with the safety factors involved in the use of plutonium and the higher actinides, the more deadly elements of radioactive waste.

Renewable energy

Another approach to the problem of filling the energy gap, whenever it may appear, is to develop renewable sources of energy such as solar power, wave power, hydro-electric schemes, geo-thermal power stations, windmills, etc. Useful as these alternative power sources might be in a regional context, the estimates are that if every one of the proposals with any chance of success was put into effect, the total contribution to the energy needs of Britain would only amount to 6 per cent of the total, or at the outside 10 per cent. This leaves 90 per cent plus to be provided by fossil fuels and nuclear power. As coal, and for a time oil, will have to be used as chemical feedstocks it would seem that, unless some important new development takes place, renewable fuels can help, but will not provide the answer to the energy problem.

The wider perspective

Britain's energy prospects for the next 20 years are surprisingly good. It is what happens at the end of the century that gives cause for alarm and concern. The basic fact is that, as far as is known at present, coal is the only fossil fuel with sufficient proven reserves for several centuries' supply at current rates of consumption. Fossil fuels are terminal resources which once used cannot be replaced. Furthermore, they are the main sources of carbon and other elements used in the chemical industry. Between them the fossil fuels provide the energy needed for transport, heat and light for domestic and industrial purposes, energy to drive factory machinery and household appliances, as well as the feedstock for chemical and other industries.

The amount of energy used is related to the size of the population and its living standards. The people of the developing countries, those living in the traditional way on subsistence agriculture, not the Westernized élites in the large cities, use about 0·5 kW per man. In Europe the average consumption is 5·0 kW and in the US 10 kW per head. If the world was brought up to American standards and consumption fuelled from the world's present oil reserves, supplies would be exhausted in 5½ years. Looked at another way, if the growth in demand for energy continues to rise at about the same average annual rate as over the last 50 years, then in the last quarter of this century mankind will consume 2·4 times more energy than in all the years of previous time before 1975.

Population growth

The use of hydrocarbons to produce the energy needed to drive machinery dates from the Industrial Revolution which began in Britain in the second half of the eighteenth century. The growth in world population began about the same time but accelerated after the Second World War, when the populations of the Third World as well as those of some developed countries increased dramatically. The introduction of soap, detergents, insecticides, vaccines and antibiotics gave the developing countries the means of death control, well ahead of any impact from the campaigns to limit population growth by birth control. The result has been an increase in world population from 1950 million to 3600 million in the quarter century from 1950 to 1975. At the current rate of 2 per cent increase a year, world population will be over 3900 million by 1980 and approaching 8000 million by 2010. It may be that population control will gain momentum in the remaining years of the century, but the signs of this happening are not promising.

Population and resources

Another aspect of the impact of population on the energy problem is its global distribution. The poorer parts of the world have large rapidly growing populations which greatly outnumber those of the industrialized countries. Asia (excluding Japan) has a population of 1971 million, Africa 403 million, Latin America 208 million, and Oceania 21 million. By contrast, Europe (excluding USSR) has 511 million and

North America 234 million.* These figures underline the difficulties facing both industrialized and developing countries. The latter are clearly not going to experience any sudden breakthrough or take-off that will heave them above the subsistence line and move their living standards up towards those of the West. At the same time, the industrialized countries are finding themselves increasingly dependent on the Third World for fuel and raw materials. It was the appreciation of the meaning of this harsh fact which enabled OPEC to destroy the complacency of the industrialized countries. Other developing countries have tried to follow OPEC's lead by forming cartels to control the supply and prices of various essential commodities. So far such moves have met with only partial success and commodity prices have continued to fluctuate widely. What has to be realized is that it is not only supplies of some fuels that are nearing exhaustion. A number of minerals including copper, lead, zinc, chrome, manganese and bauxite are approaching the point of exhaustion of the more accessible supplies. Their continued use will depend on high cost lean ores and the opening up of reserves in less accessible places. While making full allowance for the effects of the substitution of new materials, and the application of new technologies to the production and use of existing raw materials, it would be foolish to devote too high a proportion of our resources to bridging all the energy gaps and expect minerals, raw materials and foodstuffs to take care of themselves.

Malthus was not wrong

What all this means is that the apparent success of the industrialized countries in raising the living standards of their people has been only partial. In its present phase it has led to a polarization of the world into industrialized and developing countries, with the Communist states looking on. Progress has depended, as it has in the Communist states, on the diversion of reserves of fuels and raw materials away from the developing countries. The economic dilemma posed by Malthus that the pressure of population on resources would become insupportable has to be restated in modern terms. Malthus, writing in the last years of the eighteenth century,† was afraid that it would not be possible to raise

* See *The World Bank Atlas*, 1976.

† *An Essay on the Principle of Population*, first published in 1798; Penguin, 1970.

food production in Britain to a level sufficient to maintain an ever increasing population. In fact, thanks to the development of world trade in agricultural products and the improvement in farming methods, what Malthus regarded as impossible has been done. In spite of imperfections, the market economies of Europe and the US have achieved levels of economic growth that have given their citizens living standards well above those of their grandparents, and of a different dimension from those of the Third World. Malthus has been proved wrong only if, looking at our own situation, we refuse to accept the fact that our living standards cannot be achieved by the world outside the OECD and some COMECON states. Economic growth Western style is not going to solve the problems of the developing world in the foreseeable future. Population curbs, increased aid from the West, freer access to the markets of the developed world, the regulation of commodity prices, and industrialization programmes are all having some effect. However, the recipe as a whole is self-defeating. World resources are under severe strain in the present situation where roughly a quarter of the world's population consumes three-quarters of its raw materials and fuels. Changing this situation by means which depend on raising Third World living standards to those of Europe, let alone America, is impossible. For the Third World to catch up would require the use of such destructive technologies that the world would not be able to meet the demands on its resources. The Third World has a resources gap rather than an energy gap. The poorer countries of Africa and Asia would use more energy if they could move on from subsistence agriculture to manufacturing. The central problem of the world economy is unequal development.

International policies

The position of the individual OECD members varies considerably. In Western Europe only Britain and Norway have the prospect of medium-term self-sufficiency. The rest are all dependent on importing oil, and their ability to do so is qualified by their balance of payments positions and the surpluses they are able to generate from overseas trade to devote to the purchase of OPEC oil. Of the five OECD countries outside Europe – the US, Canada, Australia, Japan and New Zealand – only Canada is an energy exporting country. Japan depends for 90 per cent of its energy on imports, a situation which so far it has managed to sustain on the basis of advanced technology applied with astonishing managerial skill and industrial discipline. The fact that Japan also imports most of her raw materials and foodstuffs marks her out as a special case for study in the survival stakes. Australia has large but as yet uncharted reserves of minerals, together with established coal fields and some more recent uranium and oil discoveries. The prospects of further discoveries should be sufficiently good to ensure that Australia will not be faced by serious energy shortages for the rest of the century. New Zealand, producing agricultural supplies for South-East Asia, a region of acute food shortage, is likely to be able to afford to import whatever energy she requires above the level of domestic production. The US, still with vast reserves of coal and oil, is in a different situation. The reserves on which American wealth were founded are no longer equal to sustaining the high rate of economic growth which has produced a GNP per head of $6300. Reserves which had been left undisturbed, notably the Alaskan North Slope oil deposits, are now to be exploited. The anti-pollution laws which have prevented the extensive use of coal in power stations are being re-examined, and plans made for a considerable extension of the nuclear power programme. At the same time, energy conservation, typified by the reduction in the size of automobile engines, is being practised. In a very real sense the fight to decide how the world's energy and mineral resources will be allocated will be won or lost in North America.

Of all the oil importing countries, the energy policy adopted by the United States is the most important for the rest of the world. A continued high level of oil consumption based on imports would sustain the monopoly position of OPEC, and channel a major share of world oil supplies to the United States.

The more oil that flows into the United States, the less there will be for developing and industrialized countries alike. A better understanding of these complex international relationships is essential if national differences are to be resolved.

Economics of melodrama

The only thing to emerge on the energy front from the forecasts and scenarios dealing with the last quarter of this century is that no one knows with any certainty what will happen. The reason for this is quite simply that the energy problem is not a straightforward exercise in predicting the demand for and supply of fuels. It has widespread political ramifications which affect all international relations, but particularly those between oil producing and oil importing countries. The Cape and Suez trade routes, with their string of bases and coaling stations formerly regarded as the vital life-lines of Empire, long ago gave way to pipelines and giant tankers moving the world's oil from producing countries that could not use it to industrialized countries that could. This arrangement was conducted quietly and profitably by the oil corporations, and any disturbance of it created an international situation that the industrialized countries hastened to solve.

The 1973 oil price rise was a dividing line between two periods in international trade and development, whose effects are still being assessed. Not all of them by any means were damaging to the British economy. Without the sharp rise in oil prices, high cost North Sea oil would not now be a viable proposition. On the other hand, the ending of the era of low cost energy increased inflation and pushed Britain's balance of payments further into the red. This in turn led to mounting inflationary pressures on sterling and the need to press forward with the development of North Sea oil in order to ease the overseas payments deficit. The OPEC price increase created the oil crisis and at the same time presented Britain with the possible means of surmounting it and strengthening the economy. This is the economics of melodrama with little connection with government policies or attempts at micro or macro economic adjustments. The reason why forecasting is now more hazardous than ever is that no one knows whether there are any more OPEC-type spanners to be thrown into the works.

Technological hazards

Although it is believed that world oil reserves have been fully surveyed and production reasonably accurately forecast, changes are still possible. New fields may be discovered on land or on the continental shelves. More problematical is whether oil and gas resources will be found under the deep waters of the oceans, and if so whether they can be exploited commercially and, pending decisions from UNCLOS, the UN Law of the Sea Conference, who will develop them and on whose behalf. A more likely source of supply is that ways will be found of taking a larger proportion of oil from existing fields than is possible with current technology.

In the nuclear field there are more doubts than certainties. The Flowers Report has underlined the fact that, although the fast reactor may be perfectly safe, we cannot be sure of this until a full-scale reactor has been in operation for a trial period. The policy of embarking on a large and sustained nuclear programme with a major increase in capacity of thermal reactors, and an unspecified number of fast reactors operating by 2000, in the hope that everything will be all right, is likely to run into increasing opposition and delays in the next decade. Any forecast which depends on a major contribution from nuclear energy to total energy requirements is liable to be upset. The experience of the AGR programme is recent enough to be relevant.

Which energy gap?

In spite of misgivings about the reliability of forecasts, the question 'What is to be done?' cannot go unanswered. The option of going back to Adam Smith and leaving the market to decide how much energy should be produced is no longer open. The choice today is between more or less government intervention in the market. In the field of energy, with its long lead times on investment, the high proportion of GNP involved, and the absolute necessity of maintaining supplies, *laissez-faire* is clearly out. The best that can be hoped for is that policy decisions will not be taken in order to placate those vested interests with the most political muscle, but that they will be based on a broad view of the long-term position. The interests of Britain cannot be considered in isolation, without taking account of those of our partners in the EEC and of our international obligations and responsibilities.

The main imponderable facing the decision makers is to decide when the energy gap created by the exhaustion of North Sea oil and gas will occur. Clearly coal and nuclear energy must be available to bridge the gap when it does appear. For coal as the remaining indigenous fuel the position is clear. Capacity must be developed to help eke out North Sea oil and gas as long as possible, and to save imports of oil and gas when our own supplies are no longer available. Either way, it is necessary that the *Plan for Coal* should be fully implemented according to schedule. For nuclear energy the need for increased capacity will not come until the 'nineties. This means that the necessary investment need not be embarked on until the mid 'eighties with an interval now for detailed research and development. This should enable a view of the future of the British economy to be taken standing a little further away from the brink. What must be avoided at all costs is taking panic decisions on the basis of the forecasts made without due consideration of the highly complex energy situation. North Sea oil and gas have given Britain a breathing space in which to set the economy in order and work out an energy policy for the period when these resources are no longer available. The essential first principle is to remember that breathing spaces are meant for breathing – and the deepest breaths should be taken by the policy makers.

Back to square one, unless . . .

Looking ahead, the general conclusion is that around the year 2000 we shall be back where we started, with the windfall resources of natural gas and oil exhausted. Whether we can then bridge an energy gap depends on how we have spent the intervening years. Whether we face a return to the balance of payments deficit will depend on what steps have been taken to restructure the economy. Will new sources of energy have been discovered, will fast breeder reactors be in operation, offsetting the dangers of a shortage of uranium, and will new technologies have reduced the importance of oil for air, land and sea transport? These are all very big ifs, and it requires considerable optimism to assume that only the favourable things will have happened. When we look at the nuclear energy industry and consider that it is based on research that was going on in the 'thirties with practical application to nuclear power stations in the early 'fifties, and consider the false starts and problems that have arisen since, we realize what a long time-scale is required for new sources of energy to be developed. By the year 2000, if we have not taken advantage of the windfall of North Sea oil and gas to work out and implement positive economic policies, we shall be back looking for imports of fuel which, even if they are available, we may not be able to pay for.

Part 2
The world setting

Chapter 10
The crowded quarter century

End of a system?

The system of international economic relations created by the industrialized countries immediately after the Second World War reached the end of its first phase with the oil price revolution of October 1973. The system had no great difficulty in surviving higher oil prices, with a flurry of practical activity going on behind a barrage of protest. Indeed most of the industrialized countries coped extremely well with the problem of raising the extra money to pay their oil import bills. From an export deficit of $41 billion with the oil exporting states and a surplus of $6 billion with the non-oil developing countries, in the first half of 1974 they moved to a deficit of $21 billion with the oil exporters exactly matched by a $21 billion surplus with the rest of the developing world in 1975. Although this shift in trade and resources was not the same for all industrialized countries, the principal sufferers from the OPEC initiative were the poorer developing countries without oil resources. The higher oil prices which the international oil corporations were forced to accept were only the outward sign of what had happened. In fact, OPEC had managed to secure control of a major sector of the world economic system which had previously been the sole preserve of the rich nations.

The question now was whether the international system could stand up to the pressures likely to occur in the years ahead. The Bretton Woods institutions, while not discredited or discarded, have had to face heavy criticism and adapt to a world in which the variable factors are more numerous than they used to be. The days are gone when the annual World Bank and IMF meetings with their bland summaries of the past year's activities and placid forecasts of what lay ahead, were the highlight of the international year. From August 1971 when dollar convertibility and, with it, fixed exchange rates ended, the existing global institutions have come under attack from various rival bodies that have come into being, without making any effective rejoinder.

The fatal flaw

What OPEC has done goes far beyond the business of producing and distributing oil. It has revealed the frailty of the arrangements covering international relationships, especially those between developing and industrialized countries. The Bretton Woods institutions have been shown to have a fatal flaw, namely that the Third World had no hand in their creation. Back in 1945 when their constitutions were being drafted and voted into existence, most of the developing countries were still a part of the colonial system. In Britain, the newly elected Labour Government had not only to face up to the task of reconstruction at home, but also to inaugurate what proved to be the post-imperial situation. In the late 'forties the independent states of India, Pakistan, Sri Lanka and Burma were created and independence for Commonwealth African, Pacific and Caribbean states followed over the next twenty or so years. French, Dutch and Belgian territories achieved independence in the same period, but by different routes.

Not what it used to be

The number of independent Commonwealth countries, including Britain, is now over 30. While it would be true to say that close links still exist between the developing countries of the Commonwealth and Britain, the Commonwealth connection is not the dominant influence in determining the policies of member governments. There is no Commonwealth lobby at meetings of the Group of 77, nor are Commonwealth views pressed at UNCTAD and other international gatherings. The truth of the matter is that the Commonwealth, built up on free trade and the gold standard in the nineteenth century and revived by Imperial (later Commonwealth) Preference and the Sterling Area in the inter-war years, fell apart once the internal discipline of a common system of trade and payments was relaxed.

With virtually every member country having to be treated as a special case, considerable flexibility in finding constitutional solutions has been required. This in itself, while admirable from the legal point of view, has emphasized the differences between member countries and underlined the very tenuous nature of the forces binding the Commonwealth together. While the tactics of decolonization may have faltered from time to time, the strategy has been broadly sound and acceptable. Even so, what has emerged at the end of the day is a not very coherent group of countries joined somewhat loosely together by a common past and ties of mutual self-interest. In its new form the Commonwealth is certainly not an end in itself. The decision of the Heath Government to join the EEC, endorsed by the referendum held by the Labour Government in 1975, demonstrated the fact that Britain had not the resources to sustain the old Commonwealth relationship. This could mean that the Commonwealth will retain its hold only over cricketers and philatelists. Or it could be that, although not an end in itself, the Commonwealth could prove to be a means to an end for some, at least, of its members. The form this may take is another of the variables that will be encountered in the last quarter of this century.

Britain's continuing crisis

Any assessment of the relationship of Britain to the developing Commonwealth must take account of the connection between the performance of the British economy and the management of its relations with other countries. Throughout the third quarter of the twentieth century Britain was easing off its imperial coil, and looking for an alternative. The 'special relationship' with the US had involved the US administration in progressively taking over burdens Britain could no longer carry. It was also based on the conviction – documented by experience – that Washington would bail out sterling in any crisis. The 'special relationship' had two main disadvantages. The first was that it perpetuated the illusion of great power status which Britain had clung to from the end of the Second World War. This had its special manifestation in the belief that, given a specially favoured position in relation to the US, Britain could maintain an independent nuclear deterrent, which in fact she could not afford. The second disadvantage was that the 'special relationship' conflicted both with the need to restructure Britain's economy and the need to sort out her relationship with Europe. It was at the root of President de Gaulle's objections to British membership of the EEC and of his veto of the British application of January 1963. It is no accident that this came only four weeks after the Nassau Conference* at which Anglo-American nuclear cooperation was considered. The idea of the 'special relationship' took a different form in the proposals put forward for a North Atlantic Free Trade Area (NAFTA). This was not seriously considered by the US, which, as the leader of the Western powers, could not join a free trade area, which by definition would discriminate against non-members. The NAFTA proposal faded on the appearance of Mr Heath as Prime Minister in 1970 with a commitment to take Britain into the EEC.

Change of direction

The major event in changing the direction of economic policy in the US and Western Europe was the American declaration of dollar non-convertibility in August 1971. In international terms it was an announcement that the rules relating to trade and payments were about to be changed. The intention appeared to be to move away from direction by international institutions, governments, and central banks to an era in which the market mechanism would be given a freer run. The Smithsonian agreement in December 1971 which set the seal on the American decision provided for a new range of parities, which in fact only lasted for a few months before the world's currencies began one by one to float away. All kinds of benefits† were expected to follow. Balance of payments deficits would be taken care of by adjustment of the exchange rate without deflating the economy or handing over the national reserves. Flexible exchange rates would make incomes policies unnecessary. The fact that things have not worked out this way has been blamed on the fact that governments could not get rid of the belief that they had to be doing something. What appeared to purists to be interference in the free working of the market system was, to governments of all complexions, no more than carrying out the usually contradictory pledges made to the electorate.

* George W. Ball, *The Discipline of Power. Essentials of a Modern World Structure,* Bodley Head, London, 1968.

† See Andrew Shonfield's introduction to *International Economic Relations of the Western World 1959–71,* Vol. 2, 'International Monetary Relations', OUP for Royal Institute of International Affairs, 1976.

Chancellors of the Exchequer, whether well informed or well intentioned, characteristics which do not always go together, have difficulty in translating political aims into coherent economic objectives. At Treasury level confusion arises on whether economics is a branch of politics or vice versa. In a free-wheeling system where the market decides the long-term value of currencies, the temptation is always to denounce others when the market produces unwelcome answers, and condemn developments as the work of speculators who have miscalculated the position. A free market in international currencies operated on agreed principles would make the holding of national reserves unnecessary. Unfortunately in Britain, compulsive governmental interference in favour of the pound, known inelegantly as 'dirty floating', has ensured that national reserves will always be needed for intervention, and not simply preserved as a boost to the national ego, like the Crown jewels or the Aintree racecourse.

Not only are there too many variables, but the patterns of variation are different, and the rate of change more rapid. With the best will in the world, governments cannot leave the value of their currencies to be fixed by external pressures if to do so conflicts with domestic objectives. This was recognized in June 1976 when the Group of Ten made available a stand-by credit of $5 billion to prop up the sterling exchange rate and so prevent import prices rising to such a level that the inflation rate would have made it impossible to keep within the government's agreed limits of wage increases. In fact this was not enough, and support from the European banks for the protection of the sterling balances and a loan from the IMF followed before the end of the year.

International currency management

One of the major objectives in international economic relations in the crowded last quarter of the century must be to find an acceptable system for determining exchange rates. Neither the fixed parities of the pre-1971 IMF system, nor the 'floating' that followed have produced a mechanism that is at once reliable, effective and acceptable. Some system of international currency management has to be devised that will avoid both the inflexibility of fixed parities and the unpleasant fact that sinking follows floating. While the market can give an indication of the views of foreigners on the short-term value of a currency, it is, after all, a subjective view influenced by the market dealers' knowledge and feelings about Britain and its inhabitants, coloured by information on the current economic and political situation. Even the most objective dealer will aim off to allow for some suspected change in government policy which may, in fact, never happen. Holders of foreign currencies are always more tolerant of religious than of economic heresy.

Third World progress

The weakness of the industrialized nations has been most obvious in relations with the Third World. Their failure to meet aid obligations agreed as members of UNCTAD greatly reduced the prospects for advancement of the developing countries. Even had the whole sums represented by the target of one per cent of GNP been forthcoming, the period of development would, as the Pearson Report* pointed out, have extended well into the twenty-first century. However, the fact that a much smaller amount was forthcoming not only pushed the prospects of development further back, but affected the form and quality of the development that actually occurred. Lack of resources has produced piecemeal changes of which the least satisfactory is the creation of an educated elite in the cities of the Third World with life styles and living standards which the rest of the population can envy but never hope to emulate.

The British dilemma

For the last 25 years Britain has been trying to find an economic policy that would meet the peculiar conditions inherited from her imperial past. The realization that Britain no longer had the resources to sustain the role of leader of the Commonwealth has not brought with it any clear idea of what she might be doing instead. The long deliberations about Britain's place in Europe, whether it should be in the EEC, in some wider organization, or simply as the leader of a group of small powers, as provided by membership of EFTA, perfectly illustrates the British dilemma. At the same time the structure which had been built up for the organization of Commonwealth relations and trade was gradually dismantled.

* *Partners for Development,* (the Report of the Commission on International Development – chairman, Lester B. Pearson), Pall Mall Press, London, 1969.

Britain eventually joined the EEC on 1 January 1973. It was hoped that active participation in the political and economic development of the EEC would enable Britain to find a role appropriate to her diminished importance in the world. Unfortunately there was a lack of agreement within the political parties and the electorate at large over the advantages of British membership. The Labour Party, then in opposition, refused to take part in the work of EEC institutions, notably the European Parliament, and adopted a resolution at its annual conference favouring a referendum on whether Britain should remain in the EEC. In the event, the referendum which took place in June 1975 gave overwhelming support to British membership. Here at last, it appeared, Britain was in a position where she would be able to take an active part in the development of Europe, and through the EEC, help to develop an active external policy for the future. In practice, the EEC has not yet gone very far in developing this external role, while Britain's economic weakness has forced her into a defensive position on major EEC issues.

Limited EEC objectives

During the first 10 years of its existence the EEC had enjoyed a period of peace and quiet in world trade during which it was able to leave its external economic problems to the international institutions supported by the US. The ending of the economic Pax Americana left the Community to face an uncertain economic future. The member states were slow to realize this fact and have reacted according to their divergent interests to the problems which have arisen. The attempt to organize the EEC currencies so that they could float together within agreed margins in the so-called 'snake' failed because the members were still pursuing individual economic policies. The agreement to go forward to economic and monetary union (EMU) by 1980 was taken without planning or forethought and quickly disappeared from the agendas of meetings of Heads of Governments. Having failed to organize its own economic environment, it might be thought that the EEC would realize the importance of joining the other major economic powers in shaping new international institutions to replace the ailing Bretton Woods organizations. The reasons for this failure to rise to events are deep seated. It has become increasingly clear that except in settled international conditions the principles of freer trade and non-discrimination on which the EEC was founded are not a sufficient basis for international policy. They assume that it is possible to create a perfect market in which all industries and services are supplied under the same conditions, and member governments adapt their institutions and practices to harmonize the arrangements under which commerce and manufacturing industry are carried on.

What has emerged, however, is that in the EEC and on the wider international scene, economic systems, far from being self-sustaining, require a great deal of management involving political arrangements covering a very wide range of subjects. Fortunately, the EEC is at last beginning to lose its naïve free-trade ideology and develop more active roles in a number of sectors of external policy. A general movement is taking place away from concentration on the removal of distortions to competition towards the formulation of new objectives of structural development and social welfare. In other words, what was conceived as a purely commercial policy in terms of freer trade and regulated competition is moving towards an economic policy designed to enable the member states to make the fullest use of their resources.

The transnational aspect

The concept of the multi-national corporation investing across frontiers and making the best use of factors of production available globally rather than nationally is already familiar. The result of these activities has been to change the pattern of international trade. In the past, governments controlled trade by dealing with other governments. In the new situation, authorities such as OPEC can control the trade of their member states, and multi-nationals can influence trade in particular products. This means that trade is no longer only on a state to state basis, but also on a government to international organization or government to multi-national corporation basis. Relationships with outside bodies affecting movements of ideas, people, technology or goods now take place across frontiers in a situation in which one element is not a state. This situation, which has been developing for some time, brings the distinction between politics and economics into sharp relief. Politics involves allocating scarce resources through a budget. Economics allocates resources through a market mechanism. OPEC represents political transnational activity at its most developed. The oil majors are the best-known example of transnational economic organization at the level of the firm.

In the remaining years of the century the development of transnational relations will become increasingly important.* The limitations of the nation state have been recognized in the formation of the EEC, EFTA and other groupings in different parts of the world. The elements in the power struggle have now changed. Power today rests increasingly on economic and industrial capability so that it is no longer possible to distinguish between wealth, in the form of resources and industry, and military power as national goals. Economic and political motives have become so closely linked that it is hard to separate one from the other. On the whole, economic factors are subjugated to political aims both in the industrial states and in the Third World. Both the governments wielding political power and the corporations exerting economic power require to control people and to command resources.

The intrusive state

The liberal economic ideal of the state holding the ring and creating a situation in which business can operate is no longer a completely tenable hypothesis. The state maintains order and hands out punishments, but it is increasingly concerned to do so on its own behalf. The private sector operates under mounting restrictions imposed by the state to further policies whose objectives are not the preservation of the open economy. The state not only directs the market through pricing policies, controls consumption through taxation, and determines the level of wages, but through the operation of the growing public sector determines the direction of investment and the range of industrial products. In Britain the argument about who benefits from the market economy and how it should be controlled stems from a preoccupation with the distribution rather than the creation of wealth. In this connection a host of questions have been asked which can only be answered by observing a centrally planned economy in operation. Is the free market a bulwark of freedom or a cause of inequality? Does the market system waste resources by producing goods that are profitable instead of those that are necessary? Is the proper role of the state to intervene in the running of the economy, to compensate for the shortcomings of the free market? Is state action doomed to failure

* See Susan Strange *et al.*, *International Affairs*, July 1976.

unless it substitutes a centrally planned system for the market economy?

World economics

These questions are also being raised about the way in which the market system operates at the international level. In particular, they are being raised by the developing countries in complaints in UNCTAD and elsewhere that the global institutions work to the advantage of the industrialized countries. Commodity price levels, the management of the monetary system, methods of developing natural resources, and the GATT trade rules are all criticized and called in question. These are not matters that can be settled by one government talking to another. They are the new world economics which is concerned with formulating and implementing rules for the operation of a closed global economy. These world problems can only be solved by reconciling and adjusting national interests. Inflation is an example of a problem with national and world connotations. Competition policy for trade, the price of oil, stable commodity prices, the control of multinational corporations, and the use of technology are similarly issues on the world as well as the domestic front.

The world economy only exists in the charters of a number of largely ineffective international institutions. There are no recognized world authorities able to secure agreed objectives, or to obtain agreement on what the objectives should be.

From experience of the outcome of all the good intentions that went into creating the Bretton Woods series of international institutions, and all the misunderstandings and criticisms that have resulted, the temptation is to back down and decide that there is no hope of effectively broadening political and economic horizons. The prospect of continuing confrontation between North and South, rich and poor, clearly contains the seeds of self-destruction. Attempts to answer questions about the role of governments in controlling the economy, or the allocation of resources between production and welfare, are difficult enough at national level. When they are put in relation to the management of international economic systems, difficulties increase.

A world wide debate on how the market system operates and what should be done to improve it was staged in the North–South dialogue of the Conference on International Economic Cooperation (CIEC) held in Paris throughout 1976 and is being continued elsewhere. The production, price, and distribution of energy is one of the key issues in the discussion. It involves reconciling and adjusting national interests and policies with those of other nations to arrive at some kind of acceptable common denominator of international interest. Inflation is an example of a problem whose solution requires national governments to introduce measures to protect their own interest, and at the same time is an issue of world politics. The same is true of the energy problem. Governments seek to increase the use of indigenous supplies, and conserve energy by using it more efficiently so as to reduce the quantities required. The relations between producing countries and consumers, the level at which the oil price is fixed, the cost of research and development on nuclear and alternative sources of energy, and other problems are issues of world as well as domestic politics. Whether these different levels of interest can be reconciled will be the real test of international cooperation in the 25 years ahead.

Finding an energy strategy

What, then, should be the basis for British energy strategy to the end of the century? The form of a British energy policy is conditioned by the fact that international problems are involved. It therefore makes sense to look for cooperation with our partners in the EEC and in the wider context of the International Energy Agency. We cannot huddle round our own fire and forbid others to come near. This does not mean that Britain must share her fuel resources with other countries, and, indeed, there are few signs of any moves in that direction. Rather it implies cooperation on those aspects of energy policy which are best pursued by common action. Of these the most obvious are joint programmes of research and development on the fast reactor and on the exploitation of alternative energy sources. Cooperation can extend beyond the strict confines of the energy field, so that help in one direction can be repaid by trade-offs in the economic, financial or commercial sectors.

In looking to the future the energy situation is divided into three periods. In the first, lasting from now until North Sea oil is coming ashore in abundance in the 'eighties, Britain will continue to be a net importer of energy. In the second, which follows on and lasts until the oil begins to run out, Britain will be self-sufficient and could have surplus oil to export. In the third period, the North Sea exercise is over,

Britain's oil and gas resources will be virtually exhausted and our own special energy gap will be opening up.

Faced with this sequence of events it is easier to see the kind of things that need doing than to agree on how to do them. There is no shortage of advisers and critics telling the government what to do. The government's problem is to get things in the right order and not mix up its priorities. President de Gaulle's remark that 'Gouverner, c'est choisir' is nowhere more true than in the field of energy. It is tempting to take a pragmatic line and decide that forecasts of resources running out are nothing new, and that always in the past further resources have been discovered. But suppose it does not happen like that and those promising areas off the Falklands or in the Mediterranean, or wherever else oil has not so far appeared, do not oblige? What stands out as clearly as next year's tax demand is the fact that fossil fuels are finite resources which are rapidly being used up. Later or sooner, in our time or in that of our children or grandchildren, the same writing is on the wall. The time is coming when fossil fuels will be run down to the point where we can no longer rely on supplies. The combination of technology, science and investment started mankind on the way to equality and greater material well-being. But an awful discontinuity looms ahead if new ways cannot be found to maintain the pace of advance. The breathing space while we have the oil must not be wasted.

Policy ingredients

The first essential is to ensure that oil, gas and coal reserves are made to last as long as possible. This is a matter of prudence to give time for the orderly introduction of increased nuclear capacity, including the fast reactor, and to increase coal producing capacity and develop alternative sources of energy. Broadly there are three options, but the long lead in investment times means that the fundamental decisions have to be taken now. The first option is to bring oil ashore only in sufficient quantities to cover home demand. As the cost of production of North Sea oil is expected to be less than the cost of imports, there will be a considerable saving in foreign exchange and the balance of payments deficit will be reduced. At the same time the depletion rate of North Sea reserves will be kept down to a relatively modest level.

The second possibility is to decide to increase production rates and export the surplus oil to earn foreign currency, while cutting imports by using indigenous supplies. The danger here is that depletion rates would be raised to a level that would greatly curtail the period during which North Sea oil is available. That is, it may cut the breathing space so that there is no time to prepare for the period of shortage that will follow. While it would be a welcome change to have a few years during which the balance of payments deficit was not part of our daily fare, other problems would no doubt arise. It could be that a decade from now we might be pondering the difficulties of operating a strong currency in a world in which most of our competitors were building up deficits to pay for imported fuel supplies. Any sizeable fall in the world price of oil would upset this arrangement.

The third option is to steer a middle course and cut imports to the minimum necessary to keep the balance of light and heavy crudes in the refineries, while limiting exports to keep down the depletion rate. Agreement on depletion rates might form part of a common energy policy for the EEC. This is broadly what the Norwegian government is aiming to do. Unfortunately the difficulties facing the British economy are likely to prevent the adoption of such an elegant solution.

Chapter 11
World energy resources, distribution and trade

The affluent years

Over the past decade and more the characteristic feature of the world economic scene has been the abundance of energy. True, it has not been there just for the taking, and the poorer fuel importing countries have found it hard to raise the foreign exchange to pay for even the minimum of supplies. Many of them have found the cost of imported fuel a major constraint on economic growth. All the same, from the early 'sixties onwards the consumption of energy of all kinds has been greater than ever before. Not only were the fossil fuels – coal, oil and natural gas – widely available, but nuclear energy had gained acceptance as a 'fourth' fuel, and it was widely assumed that it was only a matter of time before the research and development programmes going on in the US, Europe, and presumably in the USSR, would perfect a practical system for producing electricity from one or other of the various alternative sources, including nuclear fusion.

Following the Industrial Revolution, Great Britain rose to early supremacy amongst industrial nations by the application of coal fired steam power to mechanical processes. The leading industrial nations up to the end of the First World War – the US, Britain and Germany – had ample reserves of coal and iron ore. The USSR moved up into the front rank in the 'thirties, using indigenous coal and hydro-electric power. Japan, another late developer, managed to take off on the basis of a highly skilled, well disciplined labour force, using imported technology, raw materials and fuel. Although coal is still the fuel with the greatest recoverable reserves in the US, the USSR and Britain, it is no longer the principal fuel used. There has been an irresistible swing from coal to oil in all the world's industrial countries, with a corresponding increase in dependence on imported energy. The developing countries, which for the most part have insufficient reserves of fuel of any kind, have become oil importers in the years since independence, paying out their hard-earned foreign currency to the newly affluent group of developing countries, the oil exporters, now banded together in OPEC. Whereas in the past the industrial areas coincided with the coalfields on the maps, industrial development is now influenced by other factors, notably availability of labour, capital and proximity to market. Moving oil from the producing areas is generally easier than trying to establish industry near the oilfields. The emergence of multinational corporations able to apply technology and capital to whatever local resources are available has introduced a new dimension into the world economy.

The growth of oil consumption

Although coal was rapidly giving way to oil by the end of the first half of this century, the change had not greatly altered the balance of industrial power. It is significant that the oil industry developed in the US, commercial exploitation really beginning with the drilling of the Titusville well by Edwin L. Drake in Pennsylvania in 1859. The early use of oil was as kerosene for lighting and heating. As the first discoveries were in the coal producing states of the east, oil met considerable competition in local markets and found its greatest use in the western states away from the coalfields. As the centres of the oil industry moved away from the states of Pennsylvania, New York, West Virginia and Ohio following the great discoveries in Texas, California and Oklahoma, the pattern of oil use began to take on a more modern aspect. In areas where there was no competition from coal it was used for steam raising in industrial plant, on the railways and for heating. But the event that had the greatest impact on the oil industry was the invention of the automobile in the 'eighties and the creation, by the early nineteen-hundreds of the great American car industry. As cars were used increasingly in the US and Western Europe, arrangements for refining oil and supplying petrol to the main population centres had to be made. The pioneer motor manufacturers in the US and Europe quickly created a demand for oil products that grew as living standards rose making car ownership possible for new groups of consumers.

At the beginning of the twentieth century, apart from the US the only oil producers of any importance were Russia (from the Black Sea area) and Burma. The increased demand for oil in the US caused concern as consumption first outstripped local supplies and then began to grow faster than new reserves were discovered. This situation stimulated more intensive prospecting in the US, resulting in the great finds in the south-western states in the late 'twenties and early 'thirties. It also sent US oil men abroad to secure concessions in the oil reserves of foreign states, and led to the creation of the great international oil corporations. Further incentive to operate abroad had come from the Sherman Anti-trust Act which, in 1911, broke up Standard Oil into its territorial constituents of which Standard Oil of New Jersey, and Standard Oil of California are the best known. In the 'twenties the American oil corporations began to operate in the Middle East. Britain was already developing the oilfields of Persia, leaving the Arab countries, except Iraq, for other states to prospect and develop.

From 1935 onwards, in spite of the heavy demands of the war years, proved reserves rose faster than oil production. Proved reserves are measured by the quantities of oil that have been located by drilling and are estimated to be recoverable by the production systems in operation at the time, plus undrilled resources so close to the drilled areas that there is every probability that they will produce when drilled. Discoveries in the Middle East brought Saudi Arabia, Bahrain and Kuwait into the ranks of the oil producing states while, across the Atlantic, Venezuela became for a time the world's biggest oil exporter, and Canada began producing oil at the Leduc Field in the late 'forties, followed a decade later by Libya and Nigeria. In the meantime additional new oil resources had been discovered in Saudi Arabia and along the Arabian Gulf.

Oil resources

Oil, like the other fossil fuels, is a finite resource. Once it has been used up it cannot be replaced. One of the big question marks over the industry is to determine how much oil there is, and how long it is likely to last. Many and various answers are given to these questions and there is no single answer that is both simple and correct. The yardstick in judging future resources is the reserve–production ratio. This takes into account the current output in any area and comes up with a broad estimate of how long production can continue at this rate unless more oil is located. As the area of the main sedimentary rocks throughout the world has largely been surveyed, the possibility of surprise discoveries is now fairly remote. Continuing improvement in the techniques of searching for oil may bring new finds in areas already surveyed, but the possibilities for second chances of this kind are also decreasing. This is why the oil industry is devoting so much attention to the continental shelves of the oceans, with more distant prospects in the deeper waters of the seaways beyond their limits. The exploration and drilling programme carried out by the oil corporations in the North Sea would never have been undertaken if the reserve–production ratio for onshore supplies had been adequate for a significant period well beyond the end of the present century.

Taking the world as a whole, total proved reserves have kept pace with consumption for the past hundred years. The great discoveries in the Middle East brought the reserve–production ratio up into a highly favourable balance. However, reserves must not only be known to exist, they must be available to consumers generally. This may be difficult because of geographical and climatic conditions – for example, in the cases of the Alaskan North Slope field and the Russian reserves in Northern Siberia. Again, supplies may be withheld for political reasons, as the OPEC embargoes demonstrated. For the first 50 years of the oil industry's development the US was both the leading producer and the greatest consumer of oil. Today only about 10 per cent of world oil reserves are in the US while over 60 per cent are in the Middle East. Looking a step further, over 70 per cent of world oil reserves are controlled by the OPEC states.

The oil in the ground

The statement that there is much more oil left in the ground than has been taken out is true in a global sense but not for individual countries. The US passed its peak of production in 1970 and has been running down since. Venezuela and Bahrain are now at a point where reserves are being run down. The major consuming countries with no oil reserves of their own have all greatly increased their imports in the past decade. Apart from the US, no important country has ever supported a high rate of oil consumption from its own resources. By contrast it has been calculated that the USSR was only using as much oil per head as Samoa in the early 'sixties.

The fact that the patterns of oil production and consumption do not match means that vast quantities of oil are moved to the consuming countries every year. Until the mid twenties most of the world's oil exports came from Mexico and the US. In the years to the end of the Second World War Venezuela was the chief supplier. From about 1948 onwards the US became a net importer with exports falling steadily behind imports. The big importing regions, however, have continued to be Western Europe and Japan. The developing countries have not generated large imports of oil. Until the oil crisis of October 1973 there seemed no good reason why the rise in imports of oil to Western Europe should not continue. The previous decade had seen an abnormal increase in the amount of energy used in the industrialized countries. Within this increase there had been a spectacular shift in consumption to oil and natural gas and away from coal. On the production side, exploration and development had concentrated increasingly on the Middle East.

One unexpected by-product of the cheap oil era was the way in which it enabled a generation of Americans to indulge in a burst of conscience. This took the form of an attack on the oil industry on conservationist grounds which successfully prevented the building of the Alaskan pipeline until it became an indispensable part of 'Operation Independence', the programme for energy self-sufficiency introduced after the 1973 oil crisis. The coal industry was prevented from extending its sales to the power stations because of the atmospheric pollution caused by burning high sulphur coal. At the same time strip mining came under attack for devastating the countryside of a number of middle western and Appalachian states. In the recent past the way to avoid censure was to use imported oil and defer development of indigenous fuel resources. This particular path is no longer so easy to follow.

Yet another problem of the international oil industry is the fact that it involves moving millions of tons of crude oil around the world. The development of giant tankers carrying 400 000 tons and more has increased the fears of pollution on a massive scale.

World energy pattern

In considering the pattern of fuel use it is important to keep a sense of perspective. Although oil is the fuel which enters most widely into world trade, the factors affecting the volume of oil exports and imports also determine the extent to which indigenous fuels are utilized. In the 'sixties oil imports rose in the industrialized countries because oil was cheap, and there were no decisive factors favouring the use of coal or other alternatives. In the 10 years to 1975, world crude oil production rose by almost 7 per cent a year. But in 1975 world crude output fell by 3·6 per cent on the previous year. The OPEC states produced 27·1 million barrels a day, or over half the world total, which was 12 per cent below the 1974 level. In part the reductions in world production of oil applied to other fuels and resulted from the decline in demand due to the world recession and lower consumption because of the unusually mild weather. However, in addition cuts were imposed on oil imports because of the sharp rise in prices. At the same time, world production of crude fell because of conservation policies introduced by a number of OPEC member states, including Venezuela, Kuwait, Libya and Nigeria.

In the Spring of 1976 the Commission of the EEC prepared a brief for EEC delegates to the Conference on International Economic Co-operation in Paris (CIEC), which forecast world energy demand to 1985. The Commission assumed that the OECD countries would make a concentrated effort to develop their indigenous energy resources bringing the group's total energy deficit down from 38 per cent in 1974 to 25 per cent in 1985. This reduction would be largely due to the development of North Sea oil and gas by Britain and Norway, the expansion of British coal production, and the bringing on stream of Alaskan oil in the US. At the same time there would be energy savings in all the OECD states but particularly in Japan and Western Europe.

The total forecast by the European Commission is that world primary demand, including the Communist states, would reach 7·7 thousand million tons oil equivalent in 1980 and 9·9 thousand million tons oil equivalent in 1985, compared with the 6·055 thousand million tons oil equivalent used in 1974. The EEC share of world demand was forecast to remain at 15 per cent so that it would continue to be the biggest net energy importer of the OECD members. Only Canada among the OECD states was a net exporter of energy during the 'seventies.

The most important development over the past 25 years has been the increasing dependence of the industrialized states on imported energy. The EEC (Nine) together imported 19 per cent of their energy needs in 1955, 46 per cent in 1965 and 61 per cent in 1972. Japan imported 23 per cent of its needs in 1955, 65 per cent in 1965 and 87 per cent in 1972. The US, by contrast, was able to depend on indigenous resources until 1965 after which energy imports rose to around 12 per cent of the total by 1972. The USSR is almost entirely self-sufficient and has the capacity to expand production. The growing dependence of the industrialized countries on imported energy, largely oil, had made them particularly vulnerable, so that they fell an easy prey to the action taken by OPEC in October 1973.

The oil corporations

Of all the changes that have taken place as a result of the 'October revolution', the shift in the position of the oil corporations is one of the most important. At the same time it is not easy to foresee what the long-term effects of this particular change will be. The seven major oil corporations which have been involved in the oil industry from its early days are all among the largest multi-national corporations in the world. However, it is wrong to suppose that they constitute the whole of the industry. Between 1953 and 1972 more than 300 private and state controlled companies either entered the foreign oil industry for the first time or significantly increased their participation in it.* Of these, some 50 were integrated international enterprises engaged in producing, distributing and refining oil, and operating in a number of countries. Among these are the so-called 'independents', all private-sector corporations, mostly multi-nationals, whose name derives from the fact that they are not part of the group of seven major oil corporations. Some of them had secured concessions in the Middle East and elsewhere and operated in the same way as the majors. Others have no exclusive supplies of crude and are dependent on purchases from whatever source these are available. Apart from the 'independents' there is a mixed group of state controlled companies. Of these, ENI (Italy), CFP (Compagnie Française des Petroles) (France), PEMEX (Mexico), and PETROBRAS (Brazil) are among the best known.

* See *The Petroleum Economist*, October 1975, pp. 362–64.

Large reserves of coal

Although oil is the most important fuel in terms of consumption and trade, coal has by far the most abundant of the world's energy reserves. However, an active programme of development by both producers and consumers is necessary to take advantage of them. On the assumption that half the measured reserves of coal can be produced economically, these would suffice to meet demand for more than a century at present rates of consumption. The proportion of proved reserves is being constantly increased as further exploration takes place and more reliable geological information becomes available. For individual countries the proved reserves are sufficient for much longer periods, and in the case of Britain are estimated to have a life of 300 years. World coal production and consumption increased during the 'sixties but not nearly so rapidly as that of oil. Only Western Europe, where the national coal industries were being run down, showed a fall in output. By contrast, considerable increases took place in the US and USSR.

International trade in coal has not been on a large scale since the 'thirties. Some 60 per cent of the total world supply of oil goes to international trade, but only 10 per cent of world coal supplies. Even a very substantial increase in world coal trade would be unlikely to have a significant effect on the position of oil in the international energy market. Energy importing countries with indigenous coal resources can therefore hope to use these to reduce their dependence on imported oil. Those without coal resources are unlikely to step up coal imports as a substitute for oil unless long-term contracts could be negotiated to ensure security of supplies. The National Coal Board is expected to be able to raise its level of exports to the rest of the EEC as part of the expansion of production provided for under the *Plan for Coal* published in June 1974 and extended in *Plan 2000* at the end of 1976. Provision for such exports would be an integral part of common energy policy arrangements worked out between the Nine. The principal uses for coal would be in electric power stations on coastal or other sites with access to sea transport, and as a feedstock for the petrochemical industry.

Britain's strategic advantage

In considering 'proved' reserves of the major fuels it is clear that, particularly in relation to the EEC, Britain is favourably placed. Attempts at estimating future energy demand come up against questions that are unanswerable in the long term. Of these the most problematical concern the future growth of GNP in the industrialized oil importing countries. The fall in demand for energy after October 1973 was partly due to the sharp increase in oil prices, but even more to the world recession. What is not clear is whether increased economic activity will result in a revival of demand for oil, irrespective of price, or whether its premium uses only will be affected, and other fuels increasingly used for other purposes. In this connection the question of security of supply is critical. Here the major potential for expanded oil production rests with Saudi Arabia, which is capable of increasing its present rate of production of 9 million barrels a day to over 20 million barrels a day. The strength of the Saudi position arises from the relationship between this flexibility in output, the price of oil and the size of its annual revenues. Production of 4 million barrels a year would now give the same annual revenue as 20 million barrels in 1972. The incentives are all in favour of restricting production and maintaining prices at a high level. As the key producer in the Middle East, Saudi Arabia is in an ideal position to act as the regulator of OPEC supplies. Anyone setting out to design the perfect organization for controlling world oil supplies would have ended up with something very like OPEC.

Natural gas reserves

In the past decade, natural gas has become a major source of energy in Western Europe. The calculation of ultimate reserves for natural gas is more difficult than for other fuels. Natural gas occurs either associated with oil or coal deposits, or by itself, as in the southern basin of the North Sea. The amount of gas associated with oil varies widely between oilfields. Even where associated gas occurs, the geographical position of the oilfield may be such that it is far too expensive to collect the gas and move it by pipeline to terminals in areas of demand. It is even more expensive to set up liquefaction plants to enable the gas to be transported in refrigerated vessels to overseas markets. It follows, therefore, that in present circumstances only a proportion, possibly less than half, of the associated gas in the world's oilfields can be regarded as a part of world energy reserves. The major gas reserves are in Russia, China, Eastern Europe and in the Middle East. Reserves in the US are no longer adequate, as was dramatically shown in the hard winter of 1976.

Various plans are being carried out to improve distribution of gas supplies within the US and to import liquified natural gas (LNG) from the USSR and the Middle East.

Uranium reserves

Consideration of the fourth fuel – nuclear energy – raises the question of the availability of uranium, the main source of fuel for the present installed fission reactors. World uranium reserves have been estimated at various cost levels. At a cost of up to $26 per kg of uranium, the 1974 World Energy Conference published a survey setting the total supply at 0·98 million tonnes. Uranium ores at lower grades than those giving uranium at $26 per kg could yield another million tonnes of uranium at prices around $2·39 per kg. If very low grades of uranium ore could be marketed, then it is possible that reserves could run to millions of tonnes of uranium. The working of very low grade ores, however, could require the use of so much energy that extracting the uranium would not produce a net addition to world energy resources. The development of fast breeder reactors which produce more fuel than they burn is regarded as the best solution to the uranium supply problem, provided that such reactors can be safely constructed and operated.

Other energy sources

Other energy sources include shale and tar sand, found principally in the US and Canada which contain large quantities of oil. Recent experience has shown that the cost of extracting this oil from any but the highest grade of shale is extremely expensive so that these deposits are not expected to make a significant contribution to world energy reserves for the foreseeable future. The most significant renewable resource at the present time is the production of electricity from hydro-electric schemes which are now producing some 7 per cent of total world energy demand. This figure could be doubled by the year 2000 and possibly trebled by the year 2020 if existing reserves of hydro power were developed.* Solar energy could represent an important addition to world energy resources in the future, but its capital cost is likely to

* 'Energy prospects', a report prepared by the Energy Research Group, Cavendish Laboratory Cambridge for the Advisory Council on Energy Conservation, reissued June 1976.

restrict its extended use in present circumstances. If energy prices rise, then the use of solar energy for water heating and possibly air conditioning and the production of electricity in those countries with suitable climates may develop. Another renewable resource is wave power which is now being developed at Edinburgh University and elsewhere with government research grants. Apart from hydro-electric power all the various renewable resources now under consideration are not regarded as likely to make a significant contribution to world energy resources by the end of the century.

Alternative sources of energy

Various sources of energy which are alternatives to fossil fuels and nuclear fission are at different stages of development. The best-known of these unconventional sources are solar, wind, wave, tidal and geothermal power. In addition, and in a quite different category, is the research now being carried out into thermonuclear fusion by the UK AEA and the EEC Torus (Jet) Programme.

The alternative energy sources are not expected to make any immediately significant contribution to the total energy supply. In the long term, say in 50 years' time, their impact could be considerable. In the short term, however, they are in competition with the established energy sources for research and development funds. In Britain the principal research programmes in this field are carried out under the auspices of the Department of Energy with the assistance of the Energy Technology Support Unit (ETSU). The principal interest in the five main alternative fuels is as sources of power for electricity generation. So far it is true to say that none has emerged as directly competitive with nuclear power. The controversy over the safety of the fast reactor and fears that the pollution level from a nuclear capacity extended to over 20 times the present level could raise serious environmental problems must tilt the balance in favour of a higher level of research spending on the alternative fuels.

Ideally, the availability of power from alternative energy sources would ease the burden of resuming large scale OPEC oil imports as North Sea oil reserves are depleted, and also the fears raised in the Flowers Report* about the dangers of accumulating vast amounts of

* 'Nuclear power and the environment', sixth report of the Royal Commission on Environmental Pollution, HMSO, London, September 1976.

some of the most deadly substances known to man, in the course of building up a nuclear programme using the fast reactor. Whether the twin fears of economic and atomic disaster will turn political attention towards the energy that could be derived from sun, wind, waves, tides and the heat in the hot interior of the earth, remains to be seen. The growing realization that we are running out of fossil fuels and that generations to come will face, in consequence, a nuclear future, is not everywhere accepted with complacency. Certainly development based on the sun, the moon and the energy concentrated in the bowels of the earth, has a less daunting aspect than a future dependent on the mysterious forces related to the nuclear cycle.

Alternative energy prospects

Of the five main alternative sources of energy, solar and geothermal power would be used primarily to produce heat, while wind, tide and wave power would be used to generate electricity. Solar energy is much the most plentiful source of power. In the course of a year, the earth receives about 12 000 times as much energy from the sun as world total energy consumption. However, the seasonal distribution of this energy is one of the main reasons why so much energy is required from other sources. If the sun's heat was more evenly distributed, the inhabitants of the northern hemisphere would not have to apply so much of their resources to keeping warm and lighting their houses and work places during the long winter months. The main possibilities for solar energy are in domestic water and space heating. The relevant techniques are already understood but their widespread utilization is held up by high cost. This state of affairs could be overcome by mass production techniques but until demand is sufficient to make this feasible, no breakthrough can be expected. According to studies by ETSU, solar energy could, given vigorous exploitation, account for 1 per cent of present primary demand for both water and space heating by the year 2000, and for 2 and 3 per cent respectively by 2030. For this to happen, all new houses, and business and industrial premises, built from now on would have to be fitted with solar panels.

As geothermal power is available in the greatest quantities near active tectonic plate boundaries, Britain is at a disadvantage compared with some other countries. The ETSU study estimates that using the hot rocks technique, satisfactory results would be obtained in Durham and Cornwall. This technique involves drilling down into impermeable rock, cracking the rock to provide a greater area for heat transfer, pumping water into the system and returning it to the surface by a second hole. The other source of geothermal power is to make use of reservoirs formed by rain water seeping through to hot rocks. This is the origin of warm springs at Bath and elsewhere. Neither geothermal nor solar energy is suitable for generating electricity.

Tidal power has a long history of experiment, but few successes. The type of scheme regarded as most likely to work involves using the tide to fill basins which are subsequently emptied through turbines. In Britain the Bristol Channel is reckoned to be the most favourable site, but studies have also been made of the Solway Firth and Morecambe Bay. There are two main types of scheme. The first involves putting a dam across an estuary, in which turbines are installed, and generating electricity from the flow of water both while the tide is flooding and ebbing. This gives an intermittent supply with no power available near high water and low water. For the electricity authority the only saving is in the fuel cost of the stations which are off-loaded while the barrage turbines are working.

The second system involves using two basins at different levels. The high level basin is filled roughly between mid and high tide, while the lower one is emptied roughly between mid and low tide. The generating turbines, placed between the basins, are able to extract continuous power. This arrangement provides a considerable saving in generating costs as it replaces power station capacity. Because of the high capital cost, long construction period with high interest charges, tidal power schemes are only competitive if fossil fuel prices are high. The CEGB has calculated that a Bristol Channel tidal power scheme could produce 10 per cent of British power needs.* As this would be a single source free from import costs and, once completed, free from rising production costs, there is a good case for a further examination of the advantages and disadvantages of the various alternative schemes before accepting the suggestion that the only way to bridge the energy gap is through a massive expansion of nuclear capacity.

Wave power has been the subject of some 350 devices patented during the past century. One device currently in process of develop-

* See J. D. Denton *et al.*, 'The potential of natural energy resources', *CEGB Research*, No. 2, CEGB, London 1975.

ment is the 'duck' on which S. H. Salter is working at Edinburgh University, based on the rocking boom principle. A different approach is seen in the system of coupled rafts devised by Sir Christopher Cockerell at Bristol University. Both of these devices have extracted a high percentage of the energy of the waves in the form of mechanical work. The problem remains of perfecting a method of transmitting the energy from a wave machine sited several miles offshore. Calculations have been made to show that sufficient energy is available to supply about half the present British energy requirement, assuming an extracting efficiency of 25 per cent from wave installations extending along some 600 miles of coast. Wave power, unlike solar power, is at its maximum in winter, provided that the structures are not overwhelmed by Atlantic storms. At this early stage estimating costs is difficult. The ETSU estimate that electricity from wave power would be more expensive than nuclear electricity but, like tidal power, it has the advantage of cutting the fuel import bill and is free from pollution hazards. An important wave power research programme including feasibility studies is being undertaken for the Department of Energy by the CEGB and ETSU.

Wind power has moved a long way since Don Quixote. Wind-driven generators of increasing size have been built in America, of which the largest, at Sandusky, Ohio, has a capacity of 100 kw. A number of large machines were built in Britain in the 'fifties, but the programme lost impetus with the fall in oil prices and the promise of cheap nuclear electricity. The change in energy prices, anxieties over security of supplies and fears about atomic pollution, have reawakened interest in aero-generators. The basic technology has already been worked out and prototype machines could be constructed. However, the windiest areas are generally farthest away from the main consuming areas, and require the building of large, obtrusive structures on hill tops. A 100 kw(e) aero-generator is being developed jointly by the National Research Development Council (NRDC) and the electrical industry.

The alternative contribution

The five alternative sources of energy could clearly make an important contribution to total energy needs, provided a vigorous development and exploitation programme was embarked upon. The best prospects appear to be in the development of tidal, wave and wind power. In each case it is a question of deciding now on the allocation of funds and suitable R & D programmes. It is clearly futile to say that R & D funds cannot be spared for the alternative fuels as they are required for the development of the fast reactor, or the thermal reactor programme.

At the same time it would be wishful thinking to decide that alternative fuels could do more than provide a very useful share of energy needs. The ETSU has calculated that with a vigorous development programme, which means adequate R & D funds being made available, the alternative fuels could contribute between 6 and 8 per cent of total energy needs at the end of the century. On this basis, given the likely increase in energy demand, a large nuclear programme would still be required. If the energy gap is as much as 250 million tons coal equivalent, as the ETSU calculates, the alternative fuels could only provide 33 million tons coal equivalent between them. Even so, the situation is not one in which we can afford to neglect any promising options. The energy which the alternative fuels could provide would increase the indigenous supply of energy at a time when North Sea oil and gas are running out. The alternative sources of energy present problems of varying technical difficulty so that it is unlikely that resources would be available to pursue all options simultaneously. If a choice has to be made, wave and solar energy present the greatest new opportunities, the prospects for wind power are already fairly well established, leaving the Bristol Channel barrage as the single major engineering project in the alternative energy field.

Chapter 12
OPEC – the rude awakening

Oil is different

Compared with coal, the oil industry operates in an atmosphere of high drama, not unmixed with intrigue. Oil is found in exotic places, it is exploited by high-technology, capital intensive methods by some of the world's largest multi-national corporations* and a number of 'independents', some of which are controlled by extremely colourful characters.† Because of their size and the commercial and industrial importance of oil, the major oil corporations are inevitably involved in the politics of the regions in which they operate. In the Middle East the task of securing supplies of oil from the Arab countries has been left to the corporations with little backing from the US or British governments. The fact that the greater part of the world's oil is found in developing countries further complicates the already troubled setting against which the industry performs.

The oil producing countries had little in common with each other except their dependence on the international oil corporations. Over the years the corporations had discovered the oil, brought it out of the ground, fixed its price and taken control over its distribution. Politics came into this relationship from time to time – as, for example, during

* These are the famous oil majors: five Americans —Exxon (Standard New Jersey), Mobil (Standard New York), Chevron (Standard California), Texaco and Gulf; BP (British) and Shell (Anglo-Dutch). CFP (France) is sometimes included as a major.
† For example, Dr Hammer of Occidental and the late Mr Paul Getty.

the Suez crisis of 1956 and the Six Day War of 1967. For the most part, however, the oil companies saw themselves as providing a buffer between the oil producing and oil consuming governments. So far as the producing countries were concerned, the oil corporations negotiated with each government on the development of its resources but the majors, between them, controlled pricing arrangements. This situation arose from the unique characteristics of the oil industry. Unlike other products, price variations have little effect on the demand for oil as for some purposes there is no substitute for it. The technology of the industry requires that production and refining are kept going at almost any cost so that a reduction in prices has little effect in limiting production. However, the rise in oil prices since October 1973 has demonstrated that the total market for petroleum products is sensitive to price increases, and that for those products such as crude oil, for which substitutes are available, market forces operate in the normal way.

Oil taxation

By the end of the Second World War, Venezuela had become the world's principal exporter of oil and the main supplier of Europe and the US. In 1945 the radical party Accion Democratica came to power and Perez Alfonso, who became Oil Minister, argued that Venezuelan oil resources were being depleted too rapidly in order to conserve American domestic reserves. In November 1948 the Venezuelan government passed a new basic law under which it secured a fifty-fifty share in all oil profits. After the first shock, the oil corporations realized that the new law gave them a measure of security by establishing the government as their partner, and so making them less vulnerable to nationalist attacks. The idea of a fifty-fifty arrangement spread to other oil producing countries and became the standard form of contract. In the Middle East its adoption by Saudi Arabia led to another important development. The occasion for introducing the new arrangement was the negotiation of a new financial agreement between King Feisal of Saudi Arabia and Aramco. Aramco, the Arabian American Oil Company, jointly owned by Socal, Texaco, Exxon and Mobil, was set up at the end of the Second World War. In December 1950, an agreement giving fifty-fifty profit sharing with 12·5 per cent royalty payable was agreed by the Saudi Arabian government and Aramco. In order to calculate government revenues, it was necessary to have some procedure for announcing the price of crude oil. This gave rise to the practice of 'posted prices' or 'tax reference' prices, as a means of working out the government 'take' per barrel. The actual calculation of government revenue for different grades of crude is a somewhat complicated process. The cost of production is deducted from the posted price. A royalty of 12$^{1}/_{2}$ per cent of the posted price is also deducted. The balance is multiplied by the income tax rate expressed as a percentage, then the royalty is added to the income tax yield. The government share represented by the income tax yield plus the royalty is also known as the 'tax paid cost'. The companies' net share is the difference between tax paid cost and the realized price for crude.

Formation of OPEC

From this brief description of the mechanics of the posted price system it is clear that governments of producing countries do not want posted prices to fall because oil revenues per barrel depend upon them. The oil companies, on the other hand, when faced with easier market conditions as in 1959 and 1960 are anxious to reduce posted prices. It was this difference in outlook that led the producing governments to form themselves into an association to secure better terms from the oil corporations. OPEC, the Organisation of Petroleum Exporting Countries, was set up in 1960 in protest at the decision of the oil companies to reduce posted prices. The initiative was taken by Venezuela at a meeting in Baghdad in September 1960, at which Iran, Iraq, Saudi Arabia and Kuwait were present. Other oil producers, such as Algeria, Libya, Abu Dhabi, Qatar, Indonesia, Nigeria and Ecuador, subsequently joined OPEC. The object of the organization was stated simply as 'the co-ordination and unification of the petroleum policies of member countries and the determination of the best means for safeguarding their interest, individually and collectively.'* The oil producing countries were by no means a homogeneous group. They had in common dependence on oil income to finance their development and balance their budgets. Also, the fact that oil is a declining asset which cannot be replaced meant that they had to formulate and put into operation economic plans in order to have the prospect of continuing development. All of them resented the fluctuations in their incomes

* OPEC, The Statute, Article 2.

brought about by changes in posted prices under the control of the oil companies. On the other hand, those producing countries with small populations were able to build up massive reserves of foreign currency over and above the needs of their own development plans. Others with much larger populations, notably Iran and Iraq, had ambitious plans for industrial development which absorbed the greater part of their oil revenues.

Until the formation of OPEC the oil producing countries had not maintained any close contact with each other in spite of their community of interests. Why then did they come together in 1960 to form a common organization when up to that time they had apparently been content to manage their relationships with the oil companies separately? The occasion for the creation of OPEC was the reduction of posted prices by an average of 10 cents a barrel initiated by Exxon in the Middle East. This followed a reduction the previous year of 18 cents a barrel which had restored oil prices to the level they were at before the 1957 Suez crisis. The main criticism of the Exxon decision was not that it reduced the revenue of the producing countries, but that it was taken without consulting them. It has been pointed out that the publication of the Federal Trade Commission report in Washington in 1952 had provided the oil producing countries with a detailed account of the activities of the US oil corporations revealed by the anti-trust investigations.* Be that as it may, there is no doubt that the leaders of the oil producing states had discovered by 1960 that the best way to face up to the oil majors was to start an organization of their own.

OPEC's first years

OPEC was set up with headquarters in Vienna with Fuad Rouhani, an Iranian, as Secretary-General.† The immediate effect of its creation was to prevent further reductions in posted prices, and although the oil companies tried to get individual governments to give tax rebates in return for extra production, the producing countries did not give way. However, OPEC did not succeed in restoring posted prices to their previous levels and its members had yet to learn how to fix prices

* See Anthony Sampson, *The Seven Sisters*, Hodder & Stoughton, London, 1975.

† OPEC moved to Geneva in 1976 after government representatives attending an OPEC Council meeting had been hijacked.

themselves and to control production. In short, although it achieved its immediate objective of maintaining prices, its members were by no means united in confronting the oil companies. In particular, the notion of conservation of resources was impossible to apply in a situation where some individual members wanted a bigger share of the market. The position was made more precarious by the Soviet Union which released large quantities of oil on to the market, making price maintenance more difficult.

The oil companies had some success in playing off the governments of the different producing countries against each other, and in refusing to deal with OPEC direct. However, the OPEC members succeeded in securing a uniform rate of royalties for each country, to be deductible from the income tax paid to them. Under the new arrangement, the 12·5 per cent royalty was deducted as an expense prior to calculating profits before tax. Expensing royalties in this way meant higher government revenues per barrel and was introduced as a uniform system for all the OPEC countries under an agreement reached in Geneva in December 1964.

The oil producing countries, unlike the producers of such commodities as copper, coffee, cocoa, and sugar, have had the advantage of increasing demand for their product over the last 25 years. Further, they have not suffered from the fluctuations in demand and prices which have created such difficulties for the producers of raw materials, foodstuffs and minerals. They have been able to take advantage of this situation because of the ability of the major oil corporations to work together informally to control the world market. This has become more difficult in recent years on account of the growing competition from the independent companies.

The Six Day War

On 4 June 1967 the Six Day War broke out with the invasion of Egypt by Israel. President Nasser alleged that Israel was supported by the US and Britain, and the Arab states agreed to shut down the oil wells and impose a boycott. However, the upset was short-lived as the oil producing states soon discovered that they were damaging themselves far more than anyone elese. Venezuela and Iran, two non-Arab members of OPEC, were not involved in the boycott and both greatly increased their shipments of oil. For the Arab countries the boycott proved extremely expensive and it was lifted at the end of June.

The most lasting outcome of the Six Day War was the closing of the Suez Canal, which continued for eight years. This did not cause any great disruption of the trade in oil as the giant tankers now in operation were not able to use the canal in any case. A second development was the formation of OAPEC, the Organization of Arab Petroleum Exporting Countries, which met in Beirut for the first time in September 1968. The founder members were Saudi Arabia, Kuwait and Libya, and the first Secretary-General was Sheikh Zaki Yamani, the Saudi Arabian Oil Minister. It was intended that OAPEC should not become involved in politics but should look after the specific interests of the Arab oil producing states. Arrangements were made for various central services to be set up and a number of projects were approved, including a dry dock at Bahrain to serve the Arab tanker fleet. The formation of OAPEC did not reduce the importance of OPEC itself, which with its wide membership has continued to develop in importance. In many ways it would be true to say that by concentrating discussion of peculiarly Arab difficulties in a separate forum, OAPEC provided an important service for the oil producing countries generally.

Participation

In the period immediately following the Six Day War the question of participation by the host governments was raised by OPEC. The Arab governments were firmly of the view that it was not enough to have the right to tax the oil companies but that they should share in the ownership of the concessions and have eventual control of the oil. Outright nationalization was not regarded with great favour because of the danger that producing countries would find themselves excluded from world markets. This had happened to Mexico in 1938 and to Iran under Mossadeq in 1951. The key word in these discussions was 'participation'. The argument within OPEC did not proceed at all smoothly and there were considerable differences between members on how fast and how far they could go. In February 1971, Algeria nationalized 51 per cent of all French interests in her oil and in November of the same year Libya announced the taking over of all BP assets, a move which was later followed by the nationalization of the assets of the remaining companies operating in the country. The other members of OPEC were more circumspect in their approach to participation. The oil corporations meantime were coming round to the view that what mattered, particularly at a time of oil shortage, was not the ownership of the oil but the ability to buy and distribute it. For the Arab countries, 1972 was the year of decision. Three of the richest of the Gulf States, Abu Dhabi, Qatar and Kuwait signed agreements for 20 per cent ownership with the companies. Iraq, after a long and bitter argument, nationalized the Iraq Petroleum Company, of which five of the major oil companies were members, in June 1972. At the same time the French company, CFP, was allowed to take a share of the oil produced. In October 1972 a general agreement was reached between Aramco and the Saudi Arabian government under which Aramco gave up 25 per cent of the established concessions, with provision for a further surrender rising to 51 per cent in 1983. Discussion still continued on the question of the price the four oil corporations constituting Aramco should pay for the oil they were able to buy back from the Saudi Arabian government's share. Agreement was not reached until September 13 1973, when the Aramco partners accepted a buy-back price of 93 per cent of the posted price.

The introduction of participation brought about a radical change in the relationships between the corporations and host governments. The most optimistic view was that it would establish a common interest between them as both were anxious to maintain the orderly development of world markets. The fifty-fifty tax arrangement of the 'fifties had produced a similar effect in the financial relationship between the two parties and it now appeared that participation might very well remove the long-standing resentment felt by a number of governments at the power of the oil corporations. However, there was not much time for any analysis in depth of the participation issue before Arab–Israeli tensions again became unbearable and the Yom Kippur War of October 1973 broke out.

The Yom Kippur War

War began just as the delegates from the oil companies and OPEC were gathering in Vienna to renegotiate prices. The general economic situation was one in which prices could only move upwards. Inflation was well above the 12 per cent a year inflation allowance agreed at the Teheran Conference in February 1971. The OPEC members complained that the price of commodities and manufactured goods which they had to import had risen much faster than oil prices. Further, the shortage of oil had strengthened the bargaining position of OPEC and some independents were already paying as much as $5 a barrel for 'participation' crude. In the negotiations the OPEC representatives demanded various concessions which would have taken the posted price to about $5 a barrel. The oil coporations were not prepared to accede to these demands, on the grounds that they were not justified by the strength of the market. They therefore requested a two week adjournment, which the OPEC delegates led by Sheikh Yamani could not accept, and the negotiations broke down.

All this was happening while the Arab–Israeli war was going on, so that military, commercial and political arguments were mixed up together. The US was accused of supplying stores and equipment to Israel and it became apparent that an embargo would be imposed. The OPEC members meeting in Kuwait decided to raise the oil price to $5·12 a barrel which was 70 per cent more than the price agreed at the Teheran Conference of 1971. The Arab members agreed together on an immediate cut back in oil production of 5 per cent which would be increased by a further 5 per cent in each month until the Israelis withdrew from all Arab territories occupied in June 1967, and the legal rights of the Palestinian people were restored. The significance of these measures had scarcely been realized in the West before the Saudi Arabian government announced a general cutback of 10 per cent plus an embargo of all oil exports to West Germany and Holland. Saudi Arabian production was cut immediately by 20 per cent and Aramco officials, Americans to a man, found themselves in a situation where they were required to enforce an embargo the intention of which was to reverse US foreign policy. So ended the less than grand design which left the Aramco partners to deal with the Saudi Arabian government while the State Department provided Israel with materials and equipment.

The oil price crisis

The Israelis agreed to a ceasefire on 21 October, but the embargo remained in force. Its effect, coming at a time of oil shortage, was to push up the price of the stocks of 'participation' crude held by producing governments. At the same time the major oil corporations were afraid that the 'independents', without secure supplies, would bid up prices still further, and that the Japanese would be prepared to pay above $6 a barrel. On December 16 the Iranian State Oil Company (NIOC) conducted an auction in which freak bids of 12 and 17 dollars a barrel were obtained. It was now clear that OPEC was in a near monopoly position and could dictate the price which the oil corporations should pay. The embargo had revealed the weakness of the companies once their access to supplies was threatened. At a meeting of OPEC in Teheran on December 22 the whole question of prices was discussed, with Iran demanding a high price of $14 while Saudi Arabia favoured a lower one on the grounds that too high a price would create a major depression in the West. If the Saudi view had prevailed it is probable that OPEC would have broken up, with a majority of members going for a high price which, in the end, the Saudis and any other producers favouring a low price would have had to accept. In the event, a compromise was reached and OPEC remained united. The posted price of oil was more than doubled to $11·65 a barrel, and a new era in the development of the oil industry had begun. Henceforth political rather than economic considerations determined the rate of production and price of oil.

The position of the corporations

Looking back, the shift in the position of the international oil corporations is at once the most important, and the most difficult to understand, of the changes resulting from the 'October Revolution'. In the principal oil producing countries the corporations have been progressively squeezed out of their traditional role of holders of concessions which gave them virtual control over crude oil production. The takeover process has not followed a uniform pattern everywhere. Some governments are now participating in production and have an agreed percentage interest in the operation: others have set up joint ventures between oil companies and state agencies; while in Iraq, Iran and Libya state control was already complete. All this means that more and more crude oil supplies are now directly or indirectly under the control of governments. The international oil corporations are themselves less fully integrated than they were and are acting as buyers rather than producers of crude oil. As a general rule they still have access to the crude oil in the countries in which they were previously responsible for production. Some Arab governments are limiting production or at least its increase, in order to conserve supply and because they have, in any case, more money than they can usefully spend. Others with commitments to large scale industrial expansion need to sell large quantities of oil to maintain the revenue which finances their economic plans. However, none of the OPEC members has the means of disposing of the large quantities of oil now under their control. The oil corporations are still responsible for marketing the oil. The significant new element in this situation is the fact that the oil companies are no longer able to determine the level of production.

The role of OPEC

In the past the major oil companies had considerable flexibility in their operations because these were spread over a number of countries. This gave them a wide variety of sources and types of oil so that customers in all parts of the world could be supplied from whichever source was the most convenient. The embargoes introduced by OAPEC governments made it extremely difficult for the oil corporations to meet their obligations, and it is unlikely that they would be able to do so if supplies were restrained in the future. Although the corporations are being eased out of production they are still an indispensable part of the international energy situation. They are still responsible for the highly complicated network of international oil movements which shift some 30 million barrels a day along the world's trade routes. They are still operating a high proportion of the oil refineries of the world, with the consequent responsibility for matching the product mix as closely as possible to the international pattern of demand. Within the importing countries they are still responsible for organizing the delivery of finished products whether in the form of four-star petrol to the filling stations, the high octane fuel for the jets at London Airport or the storage tanks of the oil fired power stations. The OPEC countries have not all reached a point where they are able to deal direct with consumers, and the corporations still act as a buffer between them and the importers even though their functions are now more limited. The principal extensions of activity of the OPEC countries have come about through the large quantities of 'participation crude' available to them as participation arrangements have been brought into force. They now have more oil to sell themselves but it does not follow that this will be an addition to the supplies available, or that prices will fall. The era of cheap oil is over and consumer countries everywhere must adjust to a higher level of energy costs.

Public opinion and the majors

The oil corporations have been much written about in the past decade. While it is true that not all the activities of all the majors have been above criticism, the fact remains that they developed a vital natural resource without which the automobile and aeroplane would not have been possible. As a result remote parts of the world have become accessible not only to the traveller, but to the impact of Western ideas. The unique feature of the operations of the oil corporations was that they were one side of a triangle whose other sides were the producer government, and the parent government of the country in which the oil corporations had their headquarters. In the early days the corporations secured concessions from governments which were not in a position to do anything about developing their own resources, and were delighted to have a secure source of income. As time went by this part of the relationship changed, as host governments gained experience, and came to realize the importance of their oil reserves. Mexico and Venezuela were the first states to expropriate foreign corporations and take control of production, while still relying on them for technology and expertise. Other states followed suit with greater or less degrees of success. The most interesting of these was the nationalization of the Anglo-Iranian Oil Company (now BP) in 1951. In this case the oil corporations refused to handle Iranian crude so that it had no market anywhere in the world. After three years President Mossadeq lost power and the Shah returned to his throne and production and trade recommenced. Anglo-Iranian did not regain its monopoly and various US oil corporations assumed control of Iranian oil. The success of the oil majors in forcing Iranian oil off world markets was made possible by the fact that supply generally exceeded demand, and by the lack of cooperation between oil producers and their consequently weak bargaining position.

The OPEC position

In the 'sixties oil was still plentiful but the producing states, joined together in OPEC, were becoming increasingly aware of the dependence of the industrialized countries on imported oil. In 1970 OPEC was strong enough to negotiate for the first time in its own right with the oil corporations, which had previously insisted on dealing with each producer government separately. The confrontation between OPEC and the oil industry in 1973 resulted in the corporations having to accept extensive producer government participation, and nationalization of their activities. In addition taxation and other payments to host governments have increased tenfold since 1970. The changed position of the oil corporations in relation to the producer governments represents a fundamental shifting of power. The corporations are still working the oil fields and marketing oil and petroleum products throughout the world. The right of the producer governments to control oil policy has been accepted. At the same time their inability to produce and market oil themselves has been recognized and the corporations are now in the position of offering services which the host governments are only too pleased to take, in return for the guarantee of a certain amount of crude and products at a preferential price.

Superficially it would appear that the oil producers and the corporations are going on as before, with only the new headings in the account books to show that the business is under new management. In fact a much more fundamental change has occurred. In the past the oil corporations were generally free from interference by their parent governments, except in times of emergency. So long as the corporations were able to deliver oil as and when required, there appeared to be no reason why they should not be left to get on with it. In the early years of the century parent governments gave active diplomatic support to the oil corporations in securing concessions and opening up new sources of supply. In short, the corporations were regarded as agents for the supply of oil, which from time to time might become involved in foreign policy. One of the most intriguing aspects of this relationship was the way in which US oil corporations were encouraged to accept double tax agreements in the Middle East as a convenient means of increasing the flow of funds to Arab states without having to secure a decision from Congress, where sympathies were predominantly pro-Israel.

The new situation

Since 1970 the old *laissez faire* attitude to the corporations has changed. The 1973 embargo and subsequent price increases demonstrated all too clearly that the corporations were no longer in control of the situation. Japan and West Germany, which relied on the majors for supplies, were especially affected by the knowledge that they were in a situation where decisions on their vital fuel supplies were being taken by corporations over which they had no control, and whose power to fulfil their obligations had been seriously eroded. In the US the oil corporations came in for renewed criticism based on Watergate allegations of involvement in corporate political payments, and the rapid rise in their profits in 1974 and 1975. As a result demands for the further dismemberment of the corporations, which may eventually be taken seriously by Congress were made.

The North Sea diversion

In Britain the development of North Sea, oil has opened up the prospect of some years of self-sufficiency, accompanied by a change of policy in relation to the oil corporations. The creation of BNOC as the British state oil corporation with participation in all oilfields in British waters has inevitably affected the position of the British major, BP, and to a less extent Shell, the Anglo-Dutch major. The same kind of reaction has taken place across the North Sea, where the Norwegians have created a new organization, Statoil, to look after national interests, in preference to leaving these to Norsk-Hydro which, like BP, is under part-government part-private ownership.

Changing priorities

The outcome of this change in the triangular relationship between oil corporations, producer and parent governments has been to increase the priority of energy policy on national and international agendas. At the international level the OECD, of which the industrialized states are members, created the International Energy Agency (IEA) after the 1973 crisis to consider arrangements for the future supply of oil and other fuels. Those member countries without adequate indigenous fuel resources, and with no connection with the oil corporations are, not surprisingly, exerting considerable influence on the activities of the IEA. So far the EEC member countries, except France, have concentrated on securing agreement in the IEA, instead of pressing for an EEC common energy policy.

The oil crisis of October 1973 was much more than a dispute over prices and supplies. Although the embargoes were occasioned by the Arab–Israeli conflict, the confrontation broadened from the quarrel over support by Western states for Israel, to the transformation of the existing international order. The OPEC states having secured control over their own indigenous resources took a further decisive step and assumed the leadership of the Third World. The industrialized nations suddenly found themselves in an entirely novel, and far from pleasant situation. While they had quickly adjusted to a dear oil regime, the prospects of being denied supplies of fuel, foodstuffs and raw materials by the governments of the developing countries filled them with alarm. For the first time since the industrial revolution of the eighteenth century the West had lost control of a critical area of economic policy. This was not a shift in the balance of power such as had brought the US to world leadership and led to the decline of the British Empire. This time power had been lost to a group outside the ranks of the Western industrialized nations. The short-term situation of high oil prices was bearable, but what of the long-term consequences? Would the producers of commodities be able to create a similar organization? Would the industries of the West come to a halt for lack of fuel and raw materials, with a consequent fall in their living standards? Would the developed countries be reduced to a situation where their principal resource was the technological knowledge needed by the Third World countries for their development? These are long-term possibilities which will perhaps not materialize in this century. But the first steps towards the redistribution of resources and world power have been taken. The oil crisis was the sign for which the Third World was waiting. Political independence from the former colonial powers had not brought with it commercial and economic independence. The pattern of development still followed the old paths with foodstuffs, raw materials and oil moving to the industrialized states in exchange for manufactured goods. Aid programmes, well intentioned as in many cases they were, brought no solution. The oil crisis showed how the break could be made in the traditional international division of labour. OPEC, having given the sign, could do no less than assume the leadership of the Third World.

Chapter 13
Energy and the new world economic order

The end of Bretton Woods

At the end of 1975 delegates gathered in Paris for yet another international conference. They came as representatives of the three main groups to which the countries of the non-Communist world now belong. These are the industrialized countries, the developing countries or Third World, and the oil exporting states. In terms of institutions this was a meeting of representatives of the members of the OECD, the Group of 77, and OPEC. The gathering, known as the Conference on International Economic Cooperation (CIEC), held its opening sessions and then set up four working committees, covering energy, raw materials, finance and development, to carry on its work. Its task was to find a basis on which a new approach to international trade and payments could be worked out in the context of the needs and aspirations of each of the three groups. In short, the CIEC was intended as a new Bretton Woods, but with the perspective changed to give a developing country view rather than continue the existing international system.

In the 30 or so years since the Bretton Woods conference the institutions then established have all worked themselves into a state of exhaustion. The International Monetary Fund (IMF) depended on a dollar whose value was fixed in terms of gold, surrounded by soundly managed national currencies, the parities of which were fixed in terms of the US dollar. The importance of fixed exchange rates was such that governments, including the Wilson Government of 1964, went to great lengths to avoid devaluation. The system worked reasonably well so long as the US was prepared to provide a continuing supply of dollars and run a balance of payments deficit. In Britain the position was complicated by the attempt to continue operating sterling as a reserve currency in spite of the time bomb represented by the sterling balances. In August 1971 President Nixon declared that the dollar would no longer be convertible into gold, and the IMF currency system was left without visible means of support. The world's currencies soon floated to whatever level the international currency market decided was appropriate. What this was depended on judgements on the stability of national governments, their ability to operate 'sound' economic policies, their balance of payments position, and a number of other pointers, on any one of which the experts could be 180 degrees out, and often were. Under the old IMF rules the international currency dealers had to decide on the validity of governmental declarations that under no circumstances would their currencies be devalued. Under a floating system dealers are able to wait to see whether declarations of intention will be backed up by action leading to the desired result, meanwhile they carefully mark the currency down, just in case the worst happens. When, as in the case of sterling, a currency is widely used in international trade and held by foreign investors, floating becomes particularly hazardous. The parities accepted in the Smithsonian Agreement in December 1971 were short lived, and most of the major currencies were floating a few months afterwards.

The industrialized countries

The oil price revolution injected a new set of uncertainties into the world of floating currencies with its arbitrary balance of payments penalties for oil importers, and bonuses for producers. For the most part the industrialized countries were able to weather the OPEC storm by stepping up use of indigenous fuel where possible, economizing on the use of oil, and raising their earnings of foreign currency by increasing exports and invisibles. As will be argued elsewhere in this book, many of the industrialized countries were most concerned at the prospect of an interruption of supplies with a consequent set-back in energy consumption and an enforced retreat to the life styles of earlier years. In modern democratic societies such decisions are politically difficult to make and hard to sustain without social and economic upheaval. The oil price revolution was therefore a highly subjective concept which governments could interpret in ways best suited to the national interest. In the US it was used to launch 'Operation Independence', a short-lived programme designed to bring energy self-sufficiency to America in the 'eighties. The immediate effect was an upsurge of activity designed to reverse the policies based on conservationist principles, notably the acceptance of development of the Alaskan North Slope and the building of a 600-mile oil pipeline from Prudhoe Bay to Valdez. At the same time, oil exploration within the US was stepped up and a drive launched for increased use of coal, previously held up by objections to the effect of strip mining on the countryside, and the air pollution from burning coal with a high sulphur content. For Britain the rise in oil prices increased the resource value of North Sea oil and gas and encouraged exploration and development of new reserves by the coal industry. The rise in oil prices meant that the high-cost North Sea oil became commercially respectable. This gave short-term advantages to the balance of payments and to the creditworthiness of the economy. In the longer term, and obscured by the ravages of the economic crisis, is the fact that Britain with indigenous oil, gas and coal moved into a different, but not necessarily stronger, position not only in relation to the other EEC members and the US, but to the OPEC states and the Third World suppliers of commodities. Whether this position will prove to be of positive benefit or simply ease the pace of our economic decline is discussed later in this book.

The Third World

The position of the developing countries which import oil is very different. These countries are feeling their way with difficulty in a complex world in which the price of oil is only one of the problems requiring solution. As relatively small users of oil products their markets, compared with those of the US or the European countries, are not big enough to attract competing suppliers. Underdevelopment and a low intake of energy go together. It is only with the beginnings of industrialization that the demand for energy begins to rise as the movement of people from the villages to the towns gathers momentum. The rise in the urban population creates an increased demand for energy, particularly from the electricity distribution network for lighting, heating and cooking. The need for improved transport facilities and finally the setting up of new industries, particularly iron and steel and cement which are energy intensive, create a demand for a rising level of fuel imports.

The linking of growing economic activity with rising living standards and increased demand for energy is not peculiar to the developing countries. In the nineteenth century, Britain, Germany, France and the US all went through exactly the same process. At that time increased demand for fuel was associated with readily available supplies of coal. This situation continued up to the Second World War and in Britain plans for post-war recovery centred round the expansion of coal production. The growing importance of oil as a replacement for coal in the industrialized countries has been one of the major economic changes of the last 30 years. It would not have been possible without the growth of the international oil corporations and their ability and willingness to serve markets throughout the world. Compared with coal, oil has many technical advantages. It is easily transported, it is not labour intensive, and is the only source of energy for air and motor transport. The only important case where coal has a dominant position is the production of coke in blast furnaces, but this is likely to be overtaken by developments in the technology of steel making, notably direct reduction.

The oil corporations

For the developing countries, with so many calls upon their capital resources, the fact that the oil industry was organized internationally meant that supplies were available to meet their energy requirements. The oil corporations established a complex international infrastructure of pipelines, oil refineries and tankers able to move oil from points where it was produced through the refining processes to the markets of the world. For the developing countries the demand was for oil products, generally in relatively small quantities, at prices determined by the oil corporations. As a rule, joint schemes were worked out by the corporations to supply particular regions at posted prices based on the price of products at the port of loading, plus the published ocean freight rate for the actual voyage. Both rates might be quite different from the actual cost and transportation charges met by the corporation. This system was breached by oil producing countries in West Africa and Latin America, which began to refine their own crude oil, and by countries such as Brazil and India whose markets were big enough to support one or more refineries. The practice of aligning freight rates on the Caribbean ended after the Second World War when Arabian Gulf posted prices were introduced.

From what has been said it is clear that the major oil corporations performed a valuable service by supplying the developing countries with the oil products they required and could not produce for themselves. For the oil producing countries in Latin America, Africa and the Middle East the oil corporations carried out search, production, and transportation of oil, all of which required advanced technology and high capital expenditure which the host countries could not provide. With increased experience of world trade both importers and producers among the developing countries found good reasons for wishing to carry out for themselves the services formerly provided by the oil corporations.

Downstream diversification

The OPEC states are beginning to diversify downstream so that, instead of exporting crude oil, which was refined elsewhere by the major oil corporations, they are participating in, or taking full responsibility for, production, refining and, in some cases, transport. In doing so they have reversed the diversification process carried out by the industrialized countries after the Second World War when they began importing crude instead of refined products. This meant a reduced foreign currency cost for the industrialized countries, while the new diversification pattern of the oil producers means increased foreign earnings.

The OPEC states

The OPEC states are themselves developing countries which happen to have a major share in the world production of a commodity which for some purposes has no substitute. They are not, as yet, important manufacturing countries, and so do not import large quantities of raw materials, as does Britain. Also, many of them have small populations so that, apart from oil, they do not enter into international trade on a significant scale.

The pattern of operations of the OPEC states means that their relationship with both the industrialized and developing countries was radically altered by the October price rise. The role of the international oil corporations was at the centre of the change. Since the introduction of participation they are now purchasers and distributors of oil rather than producers. At the same time, technological change is also working towards closer relationships between OPEC and the developing countries. The use of giant tankers makes it cheaper to transport crude oil rather than refined products. It does not follow that every developing country will soon have its own oil refinery. In many cases the establishment of a refinery is not justified and the necessary funds can be better employed for other purposes. The example of Western Europe and the larger developing countries in encouraging domestic refineries is not necessarily one which should be followed by smaller developing countries. So far as the search for new oil resources in developing countries is concerned, the international corporations are likely to continue their present operations. The major requirement for finding oil is capital and very few developing countries have sufficiently large funds to spare for such risky ventures as the search for oil. Nor do the OPEC states have the necessary skills to undertake this work. Even Burma, with its centrally planned economy, has entrusted the search for offshore oil to foreign corporations.

The North–South dialogue

The problem of oil affects all countries to a greater or lesser degree. The producing countries themselves, which superficially appear to be doing very well out of the present situation, nevertheless have their anxieties about what happens when the oil resources run out. They are also concerned as to whether inflation will destroy the value of the currency they are receiving in exchange for their scarce resources, and whether world recession will reduce demand for oil to such an extent that prices can no longer be maintained. The developing countries are concerned because the price of oil is a major constraint on development. The industrialized states are anxious about the security of supplies, the effect of high energy prices on inflation, and, above all, about the new international relationships following the rise of OPEC.

This is why the Conference on International Economic Cooperation was necessary and why it has such significance for the future. Originally it was intended to hold a conference of oil producers and consumers, namely the OPEC and OECD countries. A preliminary energy conference was held in February 1974 in Washington DC, in which the major oil importers except France took part. It was from this meeting that the IEA emerged as a forum for the discussion of energy problems. The prospect of decisions on the world energy problem being taken without their participation led the developing countries to use the special sessions of the UN Assembly to demand a 'new world economic order'. These meetings established the fact that the poor Third World oil consumers must be brought into the energy discussion, and at the same time the OPEC states declared that they were not prepared to discuss oil production, distribution and prices without discussing the problems of other raw materials, world financial problems and development generally.

As a result of the background of special sessions in the UN and other *ad hoc* gatherings, the Paris Conference became the meeting point for the rich oil producers, the industrialized nations, and the developing countries. This last group can be divided into the more affluent states, with commodities such as copper, bauxite and cocoa to export, and the poorest states, which have no means of earning additional foreign currency to pay higher oil prices. The CIEC was not the traditional North–South dialogue between the rich on one side and the poor on the other. Rather it was a meeting to consider whether the rich, industrialized countries, the OPEC states and the developing countries could arrive at a situation in which all three groups were playing a recognized and acceptable part in a revamped international system. The CIEC, in other words, represented an attempt to build a new international system in which a reformed IMF and GATT would work alongside new institutions. There is a certain element of competition with UNCTAD in these proposals. UNCTAD has up to now been the main forum for the expression of the views of the Third World countries. The tie-in between energy and commodity problems, begun in Paris and continued in the working committees, could conflict with UNCTAD's own commodity plans. Furthermore, the CIEC is concerned with global finance, which is closely interrelated with oil and commodity prices and with aid to the developing countries, which has previously been the province of the World Bank and the Development Assistance Committee (DAC) of the OECD.

Muddled thinking on commodities

Among the key issues considered at the CIEC is what should be done about trade in commodities, which the OPEC states have linked for discussion with trade in oil. The trade of developing countries consists mainly of exports of primary products (foodstuffs and raw materials) going out, and processed and manufactured goods coming in. Furthermore, both for exports and imports the main trading partners for developing countries (except for oil) are the industrialized states. One of the major problems of development is the fact that the poor countries do not trade with each other but tend to compete in the same markets with the same broad categories of exports. This is particularly true in Africa, where neighbouring countries do far more trade with Europe or the US than with each other.

This situation is well known and has had considerable influence on discussions of the development problem. However, it has led to various generalizations and misunderstandings which have muddled thinking about trade in commodities. The most important of these over-simplifications identifies the developing countries with commodity producers. Of the main commodities, only five (cocoa, coffee, tea, jute and tin) are produced solely by developing countries. For other primary products, the share of the industrialized countries is over half of total production and in some cases considerably more. In 1972 the share of the industrialized countries in total production of various commodities was as follows: wheat 77·3 per cent; barley 87·7 per cent; sugar 53·5 per cent; cotton 54·6 per cent; wool 77·8 per cent; rubber (natural and synthetic) 61·2 per cent; copper 55·4 per cent; (not including USSR) zinc 84·2 per cent; lead 81·8 per cent; bauxite 41·7 per cent; and iron ore 75·8 per cent.

It is clear that, except for tropical products, the developing countries are neither the principal producers nor important exporters of the major commodities. What is not generally realized is that many developing countries are themselves importers of commodities, some of them making large annual purchases of foodstuffs to supplement their domestic production. All of them, as their economies become more diversified, begin to import a variety of raw materials and semi-finished products, and the extent of this trade increases with growing industrialization. The idea that the interests of the developing countries would be served by raising the prices of all commodities is clearly not true. Any policy that looks at international trade in commodities solely from the angle of the producer is not in the best interests of the developing countries.

A new world economic order

An example of the way in which international discussions ignore the realities of the situation was produced in the 'Action Programme for the Establishment of a New International Economic Order' adopted by the Sixth Special General Assembly of the United Nations (Resolution 3206S.VI) on 1 May 1974. This stated that all possible efforts should be made

> to arrive at a just and equitable relation between the prices of raw materials, primary products, semi-finished manufactured goods exported by the developing countries and the prices of raw materials, primary products, food products, manufactured and semi-manufactured goods and equipment imported by them, and to work towards the establishment of a link between the prices of the exports of developing countries and the prices of their imports from the developed countries.

This statement raises a number of fundamental problems even when it is reduced to its essentials of a consideration of the relationship between the prices of commodities and those of manufactured goods imported by developing countries from the industrialized nations. The difficulty is that like is not being compared with like, and that neither commodities nor manufactured goods are produced in a vacuum. Both are influenced by a variety of commercial factors and considerations affecting their supply and price. Problems arise because of the large number of products involved, each of which has a variety of production costs, prices, and qualities. Conditions in the producing countries vary considerably and, among other things, the value of the exporting country's currency must enter into account. Also, each product has its own specific market in which supply and demand develop separately. Some commodities can be stocked, others are perishable. Some depend on fashion and other short-term factors, others become obsolete after a short time. Some products are sold on an occasional or seasonal basis while others have regular all-the-year-round markets. In some cases, there will only be a handful of producers while other products will come from many producers in many countries. Yet another important consideration is that the relationship of commodity prices to those of manufactured goods may change from time to time to avoid persistent periods of over- or under-supply.

Indexation

One of the solutions most generally put forward to help commodity producers is that prices should be indexed so that they vary with prices of manufactured goods or, put another way, that commodity prices should keep pace with inflation. However, because of the wide variety of commodities and manufactured goods entering into international trade and the different conditions under which these are produced and marketed it is practically impossible to establish a scientific index, or to determine objectively the reference prices on which to base indexation. The best that could be achieved would be an educated guess as to what the level of prices might be, and this could be subject to all kinds of objections. The protection of prices created in such an artificial manner can lead to the adoption of restrictions on supply, so as to maintain pressure on demand. Some of the measures which have been suggested are the holding back of stocks, the regulation of production, the institution of quotas for exports and the levying of uniform export taxes.

Arrangements of this kind come up in proposals at international conferences with depressing regularity. They overlook the difficulties involved when individual producers or importers have to adapt their operations to fit regulations which have been framed to cover a very broad spectrum of situations. Regulation of production might fit in very well with the development programme of a major producing state where conditions for growing a particular crop are ideal. For marginal producers who nevertheless depend on income from the same crop grown under more difficult conditions, restrictions on production could be disastrous. The holding back of supplies and raw materials from importing nations on a large scale would depress their industry and could very well lead to a permanent fall in demand for the product involved. Even more dangerous would be the prospect of accelerating inflation through a rising cost-price spiral.

Solving the commodity problem

The basic difficulties in trying to solve the problem of the relationship between commodity prices and those of manufactured goods arise from over-simplification. Most of the solutions proposed endeavour to use price policy to solve three separate problems.

1. Commodity prices fluctuate excessively and, at times do not cover the cost of production.
2. Apart from the movements in prices of primary products which it exports, a country may find that its foreign exchange earnings do not cover an adequate share of its overseas purchases. This happened to many countries following the oil price revolution.
3. Because of the fall in the value of its currency, an exporting country may find that the buying power of its earnings falls rapidly.

These three problems have to be dealt with separately and cannot be solved by pricing policies alone. The question of price fluctuation has been dealt with through commodity agreements, but it cannot be said that these have a very high record of success. This does not mean that attempts to set up international agreements for particular commodities should not be pressed forward, but it does mean that there is no general formula which can be applied to all commodities.

Stabex and other schemes

One of the most interesting recent measures for the stabilization of prices is the provision of the Lomé Convention aimed at stabilizing export earnings. However, the Stabex system, as it is called, applies only to the 52 developing countries covered by the Convention with the EEC. The developing countries of Asia with their vast populations are all excluded from this scheme. Plans of a more ambitious type were agreed in principle at UNCTAD IV at Nairobi in May 1976. UNCTAD wanted the industrialized, developing, and oil producing countries to contribute to a joint fund for the stabilization of commodity prices. The form this should take was discussed by the CIEC committee on raw materials. The most hopeful sign for the future is that commodity producing and consuming countries are at last sitting down together to discuss and study these problems. In spite of their special interests the different groups represented all recognize the interdependence of the needs of producers and consumers of raw materials and of oil. Once this is realized it should be possible to get away from the generalizations and simplifications which identify the problems of the developing countries with those of commodity producers. In the long run, low commodity prices would endanger the future of supplies, and conversely high commodity prices would turn users to substitutes or to the reduction of their purchases with the consequent collapse of the market. Market prices are not determined by chance nor fixed only for the advantage of one or other party, but reflect an almost infinite variety of supply and demand situations. On a parallel track, the relationship of oil prices to those of commodities and manufactured goods has to be considered. The fact that the OPEC states agreed at a meeting at Bali in Indonesia to freeze oil prices on condition that UNCTAD IV, going on at the same time, accepted the joint commodity fund is an indication of the likelihood of a continuing interrelationship between prices of commodities and oil, and thence of manufactured goods.

Chapter 14
The OPEC epoch unfolds

High prices and insecurity

The decisions taken by OPEC and by OAPEC (the inner group of Arab countries) in the last months of 1973 completely disrupted the conditions of energy supply, and set the pattern for the future development of the industrialized countries. At the same time the unprecedented price increases upset the already precariously balanced economies of the poorer developing countries and made inevitable a new form of relationship between them and the oil producing states.

The problem of the industrialized countries was of a different order. The rapid growth in their economies from the 'fifties onwards has been based on boundless supplies of oil from the Middle East at prices which reflected the extremely favourable conditions of extraction there.

The position of dependence on foreign fuel, which had reached 90 per cent in the case of Japan, 64 per cent in Western Europe and 18 per cent in the US, soon proved to be sustainable. Rich nations are better able to cut down consumption than poor ones which are already consuming all they can afford. Further, the rich are in a position to convert to other fuels and step up research and development on new ones.

However, although the industrialized states have become more aware of the cost of meeting their energy requirements, this is not the whole story. Apart from economic considerations, industrialized countries would not be able to ensure national defence or maintain their independence if energy supplies were subject to arbitrary reduction or sudden embargo. The threat of embargo is a powerful bargaining counter in the hands of OAPEC for securing the political and military objectives of its members in the continuing confrontation with Israel. On the wider international front, OPEC can bring pressure to bear on the industrialized states by relating the level of oil prices to a favourable outcome on issues affecting the Third World. This tactic was emphasized at UNCTAD IV in May 1976 when the decision of the OPEC states, meeting in Bali, to maintain oil prices unchanged was stated to be conditional on the West's agreement to an integrated commodity fund to be operated by UNCTAD.

The situation at the beginning of the OPEC epoch is clear to all parties. The industrialized countries face the prospect of conditions being attached to oil supplies, or of increased prices for available supplies, which they may not find acceptable. The oil producers know that they can secure political ends so long as they maintain control of oil supplies and prices. The OAPEC states can go further and impose a selective embargo on supplies from the Middle East, and manipulate prices. The developing countries are aware that they can look increasingly to OPEC for leadership on economic issues involving the industrialized countries. This is by no means a settled relationship, however, and it is one that has not yet been put to the stern test of requiring any sacrifice of self-interest by the OPEC states. The Soviet bloc is well aware of the problems posed for the security of the Western states by the new situation and draws considerable satisfaction from the fact that its own energy resources are, subject to certain qualifications, both adequate and secure.

Agenda for the West

To counter possible threats to the security of supplies, governments of the industrialized states have increased storage capacity, placed long-term contracts, and worked out rationing systems for allocating fuel supplies should the need arise. While these measures would lessen the impact of any interruption of oil supplies, they are only part of the policy requirements of an energy importing country. They need to be complemented by measures to conserve energy and increased effort to find alternative sources of supply. Economies have begun to take effect. One year after the oil price increases, energy consumption in the OECD had fallen by 2·1 per cent while GDP fell by only 0·4 per cent.

In 1975 the corresponding adjustments were a 3·3 per cent energy reduction and a drop in GDP of 2·0 per cent. It now looks as though, for most OECD countries, economies in energy consumption will be more important in reducing dependence on oil imports than efforts made to increase energy production. This means that the OECD* which broadly comprises the industrialized, oil importing countries, will reduce its total imports through economies in fuel use and a cutback in the rate of economic growth over the next decade. Increased energy supplies from OECD sources will come from a small group of oil and coal producing countries, including the US, Canada, Australia, Norway and Britain, but not all of these supplies will be available to all 24 members. Even within this inner group with proven reserves, increases cannot be realized quickly. Lead times are long, investment requirements high, and government policies often confused over objectives and methods. Again, plans which stand up well on the drawing board may come unstuck when subjected to the harsh physical conditions under which production actually takes place. The costs of production for Canadian tar sands, long regarded as a major energy reserve for the future, revealed in practice a tendency to rise so rapidly that production had to be abandoned in some cases, and heavily subsidized in others. For the OECD (except Britain and Norway) the position is that increases in oil consumption can only be met by imports from OPEC or the USSR.

The statistics and pronouncements of the OECD, or its Committee on Energy Policy, and the IEA to which most of its members belong, present a bland reassuring picture of mythical OECD average production and consumption. They conceal the fact that averages are meaningless in terms of conditions in the individual member countries. Also they do not indicate the fact that governments have shown themselves unwilling to pool their resources to any great extent, nor that when the oil embargoes were imposed they tried to secure their own supplies, and offset the effects of the price increases, by unilateral action. International cooperation is far from instinctive in times of crisis.

OPEC and the Third World

The alarm and indignation which greeted the 1973 price increases

* See Appendix 1.

revealed a preoccupation with short-term considerations on the part of importing governments. The assertion by an Arab spokesman that the increase was a helpful gesture to preserve oil reserves for future use was not perhaps appreciated at its face value at the time. All the same it was widely recognized that it would be necessary to tap new sources of energy at some point in the early part of the twenty-first century. Whether we were approaching a crisis in supply depended very much on the assumptions made. To assume that present rates of consumption would continue for the foreseeable future meant assuming also that there would be no rise in growth rates in the industrialized countries, and that the Third World states would not increase their fuel imports, and development would continue at its current depressingly slow rate. The Club of Rome forecasts of the early 'seventies were pessimistic both as to the adequacy of reserves of oil, raw materials and minerals, and rates of growth.

Meaningless forecasts

Far from simplifying the business of forecasting the availability of future supplies, the OPEC action has added a number of new variables. Although never an exact science, forecasting had a number of basic assumptions on which to found its calculations. The Group on Energy in the OECD had some handy figures which no 'expert' could afford to be without. One set showed that the electrical energy per industrial worker in the US amounted to 25 tons of coal equivalent a year. This compared with an average of 8 tons c.e. a year in Western Europe. From this calculations were made showing how long it would take for the Europeans to catch up with present US consumption, while others indicated where the Americans would be by that time. Further complications were introduced by statistics for relative rates of population increase, and by anti-pollution and environmental preservation laws which made the use of energy more expensive or prevented the development of new resources.

The position of the Third World was calculated on the purely hypothetical basis of whether its members achieved a European or an American standard of living by the year 2000. On either basis, the quantities of oil and coal required were well above the estimated possible development of supplies currently regarded as proven. The underlying assumption that the industrialized countries would generate a rate of economic growth fast enough to finance the investment necessary to meet not only their own increasing demands for energy but also to provide for the transfer of resources to the Third World was scarcely realistic. The position was further complicated by the large gap between the supply of energy to be based on fossil fuels and the supply that would have to come from nuclear power. As this was generally forecast to involve an increase in installed nuclear generating capacity of 12 to 13 per cent a year, grave doubts existed as to whether it would be technically, commercially, and environmentally possible to implement such a massive increase which would be concentrated in the industrialized countries.

It is not necessary to look back at the forecasts and prognostications that were being made about the availability and use of energy resources right up to the 1973 oil price revolution. While warning voices were raised about the growing dependence on imported oil – and, in Britain, on the folly of running down the coal industry – no one seriously questioned the ability of the major oil corporations to maintain control of the situation and continue to provide oil wherever demand existed throughout the world. The fact that the Arabs were assumed to be incapable of exploiting their own oilfields, given the large and miscellaneous array of independent oil corporations which were lining up for the chance to help them in return for the right to take up supplies, shows remarkable faith in the power of the oil majors and complete ignorance of the high qualities of the Arab people. However, when we reflect that when the Suez Canal was taken over in 1957 it was assumed that Egyptian pilots would not be able to steer a course along that remarkably straight waterway, the reaction to the OPEC initiative falls into place.

Third World institutions

The growth of Third World institutions has been so gradual that it is possible to believe that the Bretton Woods organizations still dominate the international scene. The early conferences between the developing countries were concerned primarily with questions of political independence. The African leaders came together in various Pan-African congresses to foster the independence movement in the continent. In Asia the 1955 Bandung conference raised a concept of 'positive neutralism' which was to be a 'third way', steering a course between the

Russian and American super powers. More recently, specific economic problems brought groups of Third World countries together to discuss ways and means of increasing their influence in the world economy. OPEC began in this way, and has proved to be, in commercial terms, much the most successful Third World institution. Formed in 1960 to resist further price cuts by the oil companies, its membership has now grown to 13. In the 'sixties its task was extremely difficult, as oil production was greatly in excess of world demand. In this position OPEC's main object was to ensure that the oil companies did not try to cut back payments to the producing governments. In this it was successful and by 1965 had won the right to be notified in advance of any price changes that the major oil producers proposed. By 1970 the international situation had changed and first Iran and then Libya successfully challenged the oil corporations, and their successes created a new pattern of relationships. By 1971 OPEC had reached the position where it bargained with the oil corporations on behalf of its members, instead of the former piecemeal negotiations between individual governments and corporations. In 1968, following the 1967 Arab–Israeli war, the Arab countries joined together in OAPEC (Organisation of Arab Petroleum Exporting Countries). At its creation the members were Saudi Arabia, Kuwait and Libya but these have since been joined by other Arab countries.* Although it was generally believed that the Arab states would not be able to sink their long-standing differences they have been quick to demonstrate that they have learnt the language of priorities. Forecasts that OAPEC solidarity will not stand up to realities of competition have proved unfounded.

OPEC as prototype

Attempts by other Third World countries to form similar organizations have met with much less success. The copper producing states joined together in CIPEC (Conseil Intergouvernemental des Pays Exportateurs de Cuivre), set up at the initiative of President Kaunda of Zambia at a conference of representatives of copper producing states held in Lusaka in June 1967. Its members were Zambia, Zaire, Chile and Peru. CIPEC consists of a conference of ministers, a governing

* See Appendix 1.

board and a copper information bureau. Its headquarters were established in Paris and an organization was built up centred on the development of the copper information bureau. This was in practice a statistical study group concerned with the collection and interpretation of information about the copper industry. CIPEC's efforts to control world copper prices have not met with any great success. Since 1967 the London Metal exchange price of copper has ranged from £360 to over £1360 per ton representing a greater fluctuation than at any time in the history of the copper industry. When CIPEC was formed copper prices were at a low level, about £370 per ton. They rose to £727 per ton in February 1968 but fell again to £400 per ton in 1970. After this they were generally depressed through 1971-72, with the big rise coming in 1973 up to well over £1000 per ton. In 1975-76 copper prices were depressed once more. So far CIPEC has not arrived at a workable plan for controlling the supply and therefore the price of copper. There are a number of factors preventing effective concerted action by the members of CIPEC. One of these is the different production costs in the producing countries which means that what would be regarded as a low price in one would still show a fair profit in another. Again, while copper is important in the economies of all the member countries, some of them are more dependent on it than others. While a cutback in production might be to the advantage of CIPEC as a whole, the resulting unemployment and loss of foreign exchange in government revenue might be disastrous for some of the member countries.

Although CIPEC has not been able to control production or prices, it is not without influence. It has been the means of transferring Latin American experience to Africa, and the 'Chileanization' formula for control of the copper industry has been applied in Zambia. Perhaps the greatest difficulty facing CIPEC is that, paradoxically, state-owned industries are less likely to conform to international agreements than those in the private sector. Under government ownership it may be decided a new mine would have a beneficial effect on employment, foreign exchange, or secondary industries. In such a case a government might decide to go ahead in spite of international undertakings to limit production. This kind of break-away could lead to a position of surplus and consequently lower copper prices.*

* See Sir Ronald Prain, *Copper. The Anatomy of an Industry*, Mining Journal Books, London, 1975.

Economic nationalism

The attempts to set up associations of producers, although so far relatively unsuccessful except in the case of OPEC, are likely to continue. They are a part of the struggle of the developing countries to be masters in their own houses, and, while needing investment capital, to resist a situation in which the control of development is in foreign hands. They have discovered, that it is not enough for individual governments to nationalize or secure participation in multi-national corporations operating within their own territories. They may find that the export markets for the products of these enterprises will be satisfied by the same multi-nationals from their operations in other countries. However, one major change over the last quarter century has been the decline in the power of the multi-nationals to enforce their own market decisions. In the early 'fifties, the Anglo-Iranian oil corporation was able to prevent the sale of any oil from its expropriated installations in Iran. This kind of action is no longer possible. The new situation has been influenced by two factors. The first is the emergence of a number of new countries among the ranks of foreign investors. These include Japan, Italy, and the Scandinavian countries, all of which have been fighting for a share of the markets previously dominated by US, British, and to a lesser extent, French, enterprise. In the oil industry the emergence of the Italian ENI and the Japanese Arabian Oil Company, introduced complications in the pattern of Middle East investment. The other change is the growth in the number of corporations investing in the developing countries. In the oil industry these are the 'independents', so called because they are not part of the seven major oil corporations. The appearance of state trading corporations operating in the oil industry and in commodities has introduced further complications. Among the state-owned oil entities the best-known are Petrobras of Brazil, Sonatrach of Algeria, the National Iranian Oil Company and a number of comparative newcomers which have appeared since the introduction of increased participation in the oil producing countries. These concerns are in some cases operated by personnel of the oil corporations whose assets have been taken over. In others management is supplied by companies in the smaller countries, particularly Norway, Sweden and Holland, or by Soviet bloc countries which have come forward with offers of plant and technical assistance. All this means that the pattern of investment in developing countries which emerged in the early days of independence of the 'fifties has changed. The multi-national corporations which were then the principal agents for introducing capital and transferring technology are no longer everywhere in a dominant position. They now compete with a variety of joint ventures in which host governments often have a majority share, and with state trading organizations.

What is wrong with multi-nationals?

Multi-national corporations are criticized as the manifestation of the 'unacceptable face of capitalism'. In many parts of the developing world they have become extremely vulnerable, and their actions tend to be misinterpreted whatever they do. In the years following independence, developing countries welcomed virtually all investment, whatever its source or objective. Later, demand arose for increased government control over industry. Today national plans not only call for more technologically advanced industries to be developed, but political pressures are exerted for them to be under state control. The more advanced Third World countries, such as Mexico, have been able to exclude foreign investment from a number of industries, notably electrical power, life insurance, railways and telecommunications and, more recently, oil and petrochemicals. In addition the Mexican government has already secured, or will secure in the near future, holdings in mining, road transport, fishing, shipping and bottling of soft drinks.* In other countries the list of industries completely or partly under state control could be much smaller. It seems to be the case that governments will no longer tolerate a situation in which an important industry is wholly owned and controlled by a foreign corporation. This does not mean that outright nationalization will be applied to all foreign corporations with a dominant position in the economy of a developing country. Methods vary but the intention to avoid such dominance is everywhere clear. In some cases the action of foreign investors has run counter to their commercial advantage. Many developing countries have a market which, in terms of purchasing power, cannot support more than one large supplier. However, with governments offering investment incentives on a large scale it has become the exception for a single multi-national to have a market all to itself. Foreign investors

* Louis Turner, *Multi-National Companies and the Third World*, Allen Lane, London, 1973.

have imported their own competition into developing countries with a consequent fragmentation of the market, making it difficult for adequate profit margins to be secured. At the same time, foreign corporations competing against each other are much less of a threat to the interests of the host government than a single giant in a monopoly position. The car industry in Latin America is a case in point. In 1964 Chile had 22 companies which together manufactured 7800 vehicles, each of which cost about four times as much as similar imported products. The same sort of situation developed in Mexico and Peru. In this position the host government has to endeavour to cut down the number of firms rather than try to take them all into public ownership.

Profit and remittances

The major complaint against the multi-nationals concerns the question of profits and their remittance to the investing country. Foreign firms generally aim at a return of around 15 per cent on the capital they invest, which often seems exorbitant to the host government. Clearly, if private foreign investment in the developing countries is to be forthcoming in sufficient quantity then investors and host governments must think and operate on the same wavelength. Long-standing opinions and attitudes based on experience of traditional types of investment in minerals and plantation agriculture continue to distort the picture. Years of discussion in UNCTAD and elsewhere have not succeeded in reconciling the positions of host governments and foreign investors. The former insist that private investment should be of permanent benefit to the host country, subject to nationally defined priorities and within the framework of national development plans. If it fulfils these conditions it may be encouraged by incentives and guarantees.* In contrast, overseas investors maintain that they must secure an adequate monetary return and, among other conditions, be sure of receiving compensation in the event of their assets being expropriated. The lack of understanding on this point is typified by the government of a developing country which listed among its incentives to foreign investors 'no nationalization before the end of the second year'. The difficulty is that the sort of operation which will produce a high return in a

* See the Charter of Algiers, issued by the Group of 77 in October 1967, as a preliminary to UNCTAD II.

developing country may not be within the framework of national planning priorities.

It is unfortunate that much of the discussion on the impact of foreign investment has been couched in emotional language and clouded by misunderstandings. This is particularly the case when attempts are made to measure the balance of payments impact of foreign investment by comparing the inflow of new capital with the total profits of the accumulated foreign investment in the country. Such comparisons neglect the effect of the re-investment of profits by foreign investors in the host country and the impact of foreign corporations on export promotion and import saving. Very rarely do host governments, in considering foreign investment in their territories, ask themselves what the position would have been if the investment had not taken place. The oil industry is the outstanding example of this where the oil would be unsaleable unless the corporations had invested in refineries and distribution systems throughout the world.

The 'development relationship'

While it is true that development is above all a social process, it is one which depends entirely for its success on economic progress. The 'development relationship' which is another way of describing the economic relationships between the industrialized and the developing countries, must be based on a clear division of responsibilities which meets the needs of both. After 30 years of UN and bi-lateral aid policies paralleled by a steady flow of private foreign investment, any absolute solution to the development problem is as far away as ever. International support for development aid has fallen for a variety of reasons, but increasingly because of the economic recession and the inability of industrialized countries to generate surplus funds for distribution under aid programmes. At the same time, the rise in oil prices has greatly increased the difficulties of many developing countries. While it is widely believed that a rise in commodity prices would help developing countries more than increased aid, this is by no means the case. On the contrary, the burden of high commodity prices falls most heavily on those developing countries which have to import foodstuffs and raw materials so that benefits are unequally shared.

The question of the contribution which multi-national corporations can make to the development process is much too complicated to be reduced to political slogans. What has happened to the oil corporations will not be repeated, at least in the same form, with multi-nationals in the commodity and consumer goods fields. Competition in their case is likely to come from state enterprises, national development banks, and government research agencies. At the present time the multi-national corporation is a highly effective mechanism for transferring technology from the industrialized to the developing countries. Their technical expertise and managerial know-how, together with the ability to switch resources from one part of the world to another and to market products internationally, cannot be acquired overnight by Third World governments. If the multi-nationals are replaced, the progress of development would be slowed down and the price of their departure would be a greatly reduced rate of growth. It can be argued that this in itself would be a challenge to the developing countries to find other patterns of economic activity based possibly on intermediate technology, which would give them a society more in keeping with their resources, social customs and traditions. So far the concept of intermediate technology has not made any really significant progress in practice. This is partly because the process of development over the last three decades has produced elites in the developing countries on which the benefits of such development as has taken place have been concentrated. It has been said that the whole process of aid has consisted of the transfer of money from poor people in rich countries to rich people in poor countries. The belief that this is a true statement of the position is becoming a major obstacle to development.

Epilogue

Chapter 15
1984–2000

No help from history

It used to be said that the only thing to be learnt from history is that we do not learn from history. It could be argued that the only thing to be learnt from energy forecasts is that, while all are likely to be wrong, some will be farther off the mark than others. While accepting both these propositions, it is still possible to argue the necessity for a long-term energy policy. The most compelling reason is that energy is concerned with finite resources, so that some action must be taken as they reach different stages of exhaustion. Economic progress measured in terms of rising living standards based on faster growth inevitably means more rapid depletion of fuel resources. If the rich countries continue indefinitely at their present level of energy consumption the known reserves of fossil fuels would not be sufficient for the countries of the Third World to attain European or American standards of living in any foreseeable future. A rapid increase in economic growth rates for the developing countries over the next decade would come up against the constraint of shortage of energy.

Moving from the general to the particular, the prospects for Britain for the next 20 or so years are not by any means unfavourable. This does not mean that Britain is about to return to its earlier pre-eminence as an international investor and provider of industrial goods. Too much has happened in Britain and the rest of the world for a resumption of the role of workshop of the world and principal international banker, with investment and manufactured goods flowing out, and foodstuffs and raw materials flowing in to these islands. Those were the days when Adam Smith's vision of a free, self-regulating, wealth creating, world-wide commercial system came nearest to being a reality. The market system could be seen in operation providing an ever increasing supply of goods and raw materials in response to the effective demand of the world trading nations.

The US is different

With the spread of industrialization and the opening up of the American West the British economy proved too small to sustain a worldwide role. The eventual collapse of sterling as an international currency left a gaping hole in the world economic system. The lack of equilibrium in economic relationships, due to the disproportionate wealth of the industrialized countries compared with the developing world, represents a continuing threat to peace and to constructive development. The emergence of new communist states has further increased the tension in an already difficult situation. As the Commonwealth gradually declined, the US did not, as might have been expected, naturally and automatically take over Britain's international role. One reason for this was that US exports consisted not only of manufactures but also of raw materials. The US home market was protected by tariffs so that foreign countries found it extremely difficult to sell sufficient goods to America to cover US investment and sales abroad. The result was the dollar shortage and the drain of gold across the Atlantic beginning in the inter-war years and increasing in intensity in the late 'fifties and 'sixties. With the development of new technologies based on mass production for the continental American market, the old self-regulating, self-balancing market system, of which Adam Smith wrote, no longer bore any resemblance to the actual situation. The long delay in bringing the European states together in a Common Market added to the difficulties of replacing the old basis of international trade with something more appropriate to modern conditions. The granting of independence to the colonial territories of the Commonwealth, France and Belgium and the Netherlands from the late 'forties onwards shifted the problem of development on to the world stage and added a further cause of disequilibrium.

The simplification trap

The great increase in the availability of oil supplies in the 'sixties represented an opportunity to the manufacturing and transport industries to use a new and cheaper fuel, while presenting a growing challenge to the position of coal. The development of energy policy during the post-war period demonstrated the difficulties of interpreting long-term trends in terms of short-term policies. Businessmen quite rightly dread the simplification of economic problems to the point where detailed policy calculations are reduced to short statements on which value judgements are based. The fuel industries are especially exposed to this particular danger. Much of the difficulty arises from confusion over the criteria on which investment policy should be made. Should the basis of choice between investment in coal, or nuclear energy, or natural gas or oil, be on the basis of improved financial return, technical efficiency, social considerations or the general impact on the economy? Precise financial comparisons between fuels are difficult because of governmental intervention in the energy market designed to prop up prices or maintain employment. In addition, because of the peculiar character of the fuel industries with their long investment leads, decisions which determine the power station fuel for the next 20 or 30 years can distort the pattern of both supply and demand.

The great divide between the old and the new energy situations was the oil price crisis of October 1973. Up to that time it was possible to believe that world trade and industry would flow forward on an ever-expanding supply of cheap OPEC oil. Since then it has become clear that oil will never again be cheap, that its supply is not unlimited, and that coal, nuclear energy and the so-called alternative sources of energy have all become much more important in forward projections of total energy consumption. At present it is impossible to say how much of future world demand for electricity will be met from nuclear power stations. The controversy over safety, security and pollution from the fast reactor has assumed international proportions. At the same time, although Britain's reserves of coal are sufficient for several centuries, their exploitation involves a complete restructuring of the industry. The creation of new capacity, which is necessary if the coal industry is to increase output, requires the opening up of reserves in areas where coal mining has not previously taken place. The public outcry against the 'desecration' of areas of natural beauty, such as the Vale of Belvoir, in order to develop the important coal reserves beneath its surface is in no way less vehement than that of anti-nuclear protesters.

The great fuel spectacular

In short, the energy problem is developing on the lines of a Greek tragedy, with differing groups of participants each acting out their different roles. The fuel industries are aware of the need for increased supplies of energy to meet demands for faster economic growth. The scientists and technologists know that to increase total supplies of energy will inevitably mean building more nuclear power stations, with fast reactors to ensure supplies of uranium. The politicians see the connection between the use of indigenous as opposed to imported fuel in terms of surplus or deficit in the balance of payments. In the background, as a sort of chorus, the developing countries with no fossil fuel resources of their own and no money to pay for reactors, demand help in raising the living standards of their people through industrialization. A second chorus, consisting of the OPEC oil producing states, declaims against the industrialized countries for using up fuel resources too quickly and so depriving the developing countries of the means of breaking out of their poverty. All appear to be locked together in a relationship of eternal frustration heightened by the growing awareness that supplies of oil and natural gas will soon begin to decline.

Whenever predictions of an oil scarcity have been made, new reserves were discovered, but who knows whether this fortunate circumstance is likely to be perpetuated? What is clear is that for a period of 12, 15, or more years, the UK should have a surplus of oil, to the benefit of the balance of payments position. Once this is exhausted imports of oil will begin once more, with the difference that world supplies will have become scarcer, more expensive and less dependable in the meantime so that the impact on the balance of payments will be even heavier than in the mid 'seventies. If nothing is done in the meantime to restructure the economy to provide alternative sources of energy, the British economy will not be strong enough to sustain fuel imports, at 2000 AD prices, at a level necessary to maintain living standards. Bleak as the situation of the mid 'seventies has been, what is likely to be experienced 20 years hence will be considerably worse.

What do we know?

While it is important not to attempt to read too much into economic forecasts, certain assumptions can be made. The first is that it takes about 30 years for any new fuel to be brought into operation on a large scale, and that the adaptation and extension of supply of existing fuels is also a slow process. The development of new capacity under the *Plan for Coal* is going to take a decade or more before it takes effect. The first assumption, therefore, is that changes in the total available energy at the end of the century will have been the result of changes in the pattern of existing fuel production and use rather than some technological miracle.

The next aspect of the problem on which a pronouncement can safely be made concerns the decision making process. Governments, which make key decisions on investment policies, operate on a time scale of five years, stretching from one election to the next. They tend, therefore, to be preoccupied with short-term problems which will affect electoral prospects, rather than with the situation in the time of their children and grandchildren. Although the action of a particular government in taking a major decision on a long-term project may look well in the history books, in political terms the benefits will be reaped by following governments. Electorally there is more mileage in maintaining employment for a few more months in a pit threatened with closure, or deciding to site some new installation in a marginal constituency, or making a well-timed decision on the price of electricity or gas, than in boldly opting for an enlarged nuclear programme.

To be fair, on a large number of subjects governments are powerless to take effective action. No amount of argument is likely to induce the OPEC countries that it is in their interests to reduce the price of oil. Natural prudence and the desire not to stimulate inflation and so reduce the value of money they receive for their oil may limit the size of increases, but will certainly not move prices downwards. Similarly, governments cannot be certain that technological advances will be achieved on the lines predicted by experts. A setback in the development of the fast reactor could greatly reduce the future supply of nuclear energy. A decision by the principal uranium producing countries to form an OPEC-type body (called, perhaps, URANUS) could seriously affect the price and availability of supplies in the short term. Such decisions, implemented by producers of certain minerals and raw materials, could upset the rate of economic growth in the industrialized countries. Again, governments are limited by the amount of social disruption that a particular policy is likely to cause. For example, it would be extremely difficult for a government to reduce the size of the British car industry, concentrate car production on exports, and encourage the use of public transport. Resulting unemployment in the car factories, various supply industries, and among road and motorway builders would create a reaction which few governments would care to face. In political terms the problem is to assess whether the effects of unemployment due to a lack of fuel supplies, or due to government policy aimed at reducing demand for fuel, would be the more serious. Finally, no government can influence the weather, and the disturbance it can cause to energy forecasts and policies.

Rising prices

The costs of fossil fuels, coal, oil, and natural gas, will continue to rise. This is because the more easily accessible reserves and the richer grades are exhausted first, so that exploitation moves on to lower grade, leaner supplies which inevitably have a higher production cost. Irrespective of the action of producers in ganging up to raise prices, there is an inevitable upward movement following rising fuel costs. This is why it is so important that renewable sources of energy – solar, wind, wave, etc. – should be developed, and why nuclear energy which has already reached the stage of commercial development, is so important. Conservation measures to insulate houses and factories, improve the conversion rates of fuel into electricity and other measures, would enable the same amount of fuel to satisfy a growing demand. Alternatively, with conservation a smaller amount of energy would satisfy present levels of demand. Although conservation policies are obviously the most sensible option they are not easy to introduce or, indeed, to follow. It is easier to go on driving a large car than to incur the expense and trouble of selling it and buying a small one. Higher standards of insulation for new houses are easier to achieve than trying to raise the standard of existing houses. This may increase demand for materials that may not be readily available, and raise the capital cost of houses. In conservation, as in other aspects of energy policy, decision making is a matter of choice involving conflict between long-term and short-term requirements. Governments shy away from large problems or break them down into smaller problems that they can deal with leaving the serious ones for their successors. Generally speaking, technical problems, provided adequate funds for research and development are available, are easier to solve than political ones which depend on variations in the opinions and attitudes of the mass of the population.

The small miracle

The 1973 oil crisis underlined the way in which British energy policy has always been determined by world events. By a strange chance of our national fortunes, this situation is about to be changed by the development of North Sea oil and gas. What has happened is the energy equivalent of the occasion when all the well-fancied horses in the Grand National fell at the same fence leaving an outsider who was far enough behind to avoid the pile-up, to go through and win. If the lucky jockey, faced by this unexpected change in his situation, had fallen off his horse, or into a daydream about how this would affect his future, or begun working out his tax liability, he might have been overtaken by the jockeys who managed to disentangle their mounts from the mix-up and pressed on towards the finish. There are dangers in trying to ride a metaphor too far or too hard. Suffice it to say that having emerged from the 1973 oil crisis with a chance of ample supplies of oil until the end of the century, Britain, given sensible policies, has a decided advantage over those industrialized countries without benefit of indigenous oil supplies.

Our remaining asset

Up to the end of 1976, the oil programme was adding to the balance of payments deficit. With the phasing out of payments for equipment and with production of oil running on schedule, benefits from oil would rise, according to the Treasury, from £5 800 million in 1980 to £16 000 million in 1985. The benefits from North Sea gas in the same period would go up from £4 000 million to £8 000 million. Clearly, these large sums bring problems as well as blessings. Unless used carefully, they could become a major inflationary element in the economy. This would happen if they were absorbed through the tax system to swell the proportion of GNP accounted for by public spending. There would be nothing easier than to embark on a great attack on public squalor (opposite of private affluence), in the shape of new schools, colleges, hospitals, leisure centres, municipal buildings, motorways, and so on. If this happened the position as the next century dawns, with the oil and gas from the North Sea nearing depletion, would simply be the start of the next episode in the rakes' progress, the one after the inheritance had been dissipated. A likely sequence of events would be a consumer boom based on full employment stimulated by increased public spending, higher salaries all round following rises in the public sector, and increased social security benefits for the disadvantaged. All of which would suck in imports, restore the balance of payments to its traditional deficit, and obliterate the North Sea oil surplus.

It is to be hoped that what will happen is that extra resources secured by substituting North Sea oil for imported oil and exporting surplus production, would be concentrated on the development of manufacturing industry, rather than on services and consumption. This would mean avoiding too large an appreciation of sterling, partly by rebuilding the reserves, and partly by reducing interest rates. An over-valued exchange rate could lead to a substantial deficit building up in the non-oil balance of payments. A strict control of the money supply in the 'eighties, while the oil surplus exists, will be necessary. North Sea oil is a windfall, and although not susceptible to normal economic forecasting techniques, windfalls nevertheless are affected by, and in turn affect, the working of the economic system. It has been pointed out on numerous occasions that the years of surplus should be used as a breathing space in which to restructure the British economy. If we come out on the other side without having built up industrial capacity and the level of exports of manufactured goods, we shall find ourselves having to buy oil at end of century world prices, which will be much higher than those of today. At that time there will be no great hope of borrowing from the IMF which, like banks of less degree, does not regard used up resources as good security.

To conclude, the view from the shore is still very much better than it would have been had drilling for oil and gas never taken place on the Continental Shelf. But success is flawed by the fact that the failure of British economic policies in the 'seventies has reduced the value to the economy of our new resources and they have been developed on a programme dictated by balance of payments criteria rather than by energy policy requirements. If we regard North Sea oil as our major asset, we must be prepared to realize its potential in the manner most likely to help our future development. Otherwise, we are likely to become increasingly aware that economic nationalism is a luxury we can no longer afford.

Appendices

Appendix 1
Membership line-up

Membership line-up

OECD — Organisation for Economic Cooperation and Development

Austria, Australia, Belgium, Canada, Denmark, Finland, France, Germany, Greece, Iceland, Ireland, Italy, Japan, Luxembourg, Netherlands, New Zealand, Norway, Portugal, Spain, Sweden, Switzerland, Turkey, United Kingdom, United States.

IEA — International Energy Agency

Austria, Belgium, Canada, Denmark, Germany, Greece, Ireland, Italy, Japan, Luxembourg, Netherlands, New Zealand, Norway, Spain, Sweden, Switzerland, Turkey, United Kingdom, United States.

EEC — European Economic Community

Belgium, Denmark, France, Germany, Ireland, Italy, Luxembourg, Netherlands, United Kingdom.

OPEC Organisation of Petroleum Exporting Countries

Algeria, Ecuador, Gabon, Indonesia, Iran, Iraq, Kuwait, Libya, Nigeria, Qatar, Saudi Arabia, United Arab Emirates, Venezuela.

OAPEC Organisation of Arab Petroleum Exporting Countries

Algeria, Bahrain, Egypt, Iraq, Kuwait, Libya, Qatar, Saudi Arabia, Syria, United Arab Emirates.

Appendix 2
Source material

Official publications

'Report of the Committee on the use of National Fuel and Power Resources' (the Ridley Report), Cmnd 8647, HMSO, London, 1952.

The National Plan, Department of Economic Affairs, Cmnd 2764 HMSO, London, September 1965.

'Fuel policy', Ministry of Power, Cmnd 2798, HMSO, London, October 1965.

'Fuel policy', Ministry of Power, Cmnd 3438, HMSO, London, November 1967.

'Energy conservation in the UK: achievements, aims and options', NEDO, HMSO, London, 1974.

'Energy conservation: A study by the Central Policy Review Staff', HMSO, London, 1974.

'Production and reserves of oil and gas in the UK', Department of Energy, HMSO, London, 1974.

'Coal industry examination. Interim report', Department of Energy, HMSO, London, June 1974.

'Coal industry examination. Final report', Department of Energy, HMSO, London, 1974.

'Nuclear power and the environment. Sixth report of the Royal Commission on Environmental Pollution' (the Flowers Report), HMSO, London, September 1976.

'Development of the oil and gas resources of the UK' (the Brown Book), Department of Energy, HMSO, London, published annually since 1974.

Annual reports

British Gas Corporation
Electricity Council
National Coal Board
UK Atomic Energy Council

General

Louis Turner, *Multi-National Companies and the Third World*, Allen Lane, London, 1973.

George W. Ball, *The Discipline of Power. Essentials of a Modern World Structure*, Bodley Head, London, 1968.

Andrew Shonfield (ed.), *International Economic Relations of the Western World 1959-71*, Vol. 2, 'International monetary relations', Oxford University Press for Royal Institute of International Affairs, London, 1976.

Lester B. Pearson, *Partners for Development* (the Report of the Commission on International Development; chairman, Lester B. Pearson), Pall Mall Press, London, 1969.

The United Kingdom in 1980; the Hudson Report, Associated Business Programmes, London, 1974.

Professor Meadows (ed.), *Limits to Growth (The Club of Rome)*, Penguin Books, Harmondsworth, 1972.

Barbara Ward, *Space Ship Earth* (the Pegram Lectures, Brookhaven National Laboratory), Hamish Hamilton, 1966.

WAES (Workshop on Alternative Energy Strategies), *Energy, Global Prospects 1985-2000*, McGraw Hill, London, 1977.

OECD, *World Energy Outlook*, Paris, 1977.

Barbara Ward and Rene Dubos, *Only One Earth, The Care and Maintenance of a small Planet*, Penguin, London, 1976.

Fossil fuels

Colin Robinson, 'Competition for fuel', IEA Occasional Paper 31, Institute of Economic Affairs, London.

J. E. Hartshorn, *Oil Companies and Governments*, Faber, London, 1967.

Anthony Sampson, *The Seven Sisters*, Hodder and Stoughton, London, 1975.

Michael P. Jackson, *The Price of Coal*, Croom Helm, London, 1975.

D. I. Mackay and G. A. Mackay, *The Political Economy of the North Sea*, Martin Robertson, London, 1975.

Roger Vielvoye et al. *Coal, Technology for Britain's Future*, Macmillan, London, 1976.

A. R. Griffin, *The British Coalmining Industry, Retrospect and Prospect*, Moorland Publishing, Buxton, 1976

Energy policy

'Energy prospects', A report prepared by the Energy Policy Group, Cavendish Laboratory, Cambridge, for the Advisory Council on Energy Conservation, June 1976.

ACORD, 'Energy R & D in the United Kingdom. A discussion document', ACORD (Advisory Council on Research and Development for Fuel and Power), London, June 1976.

Gerald Foley, *The Energy Question*, Penguin, London, 1976.

NEDO, 'Financial analysis of the nationalised industries', NEDO, HMSO, London, 1976.

'Towards a new energy pattern in Europe (the Robinson Report), OECD, Paris, 1960.

'Europe's growing needs for energy – how they can be met' (the Hartley Report), OECD, Paris, 1 May 1956.

Walter C. Patterson, *Nuclear Power*, Penguin, London, 1976.

Peter Chapman, *Fuel's Paradise. Energy Options for Britain*, Penguin, London, 1975.

Index

Phototypesetting in Great Britain by George Over Ltd., London and Rugby